A SYSTEMS APPROACH TO EDUCATIONAL ADMINISTRATION

A SYSTEMS APPROACH T

Robert C. Maxson
Auburn University at Montgomery

Walter E. Sistrunk
Mississippi State University

EDUCATIONAL ADMINISTRATION

WM. C. BROWN COMPANY PUBLISHERS
Dubuque, Iowa

Copyright © 1973 by Wm. C. Brown Company Publishers

Library of Congress Catalog Card Number: 72-83850

ISBN 0—697—06253—8

All rights reserved. No part of this publication may be reproduced, stored in a retrieval system, or transmitted, in any form or by any means, electronic, mechanical, photocopying, recording, or otherwise, without the prior written permission of the copyright owner.

Printed in the United States of America

This book is dedicated to Sylvia and Marian

CONTRIBUTING AUTHORS

ADAMS, DEWEY A., Professor of Education, Virginia Polytechnic Institute and State University

ATWELL, CHARLES, Assistant Dean, College of Education, Virginia Polytechnic Institute and State University

BURNS, RICHARD W., Professor of Education, University of Texas at El Paso

DEMEKE, HOWARD J., Bureau of Educational Research and Services, College of Education, Arizona State University

GARVUE, ROBERT J., Professor of Education, Florida State University

GILLILAND, JOHN W., Director, Center for Educational Facilities, College of Education, University of Florida

HOLDER, LYAL E., Associate Professor of Education, Brigham Young University

JARVIS, OSCAR T., Professor of Education, University of Texas at El Paso

JOHANSEN, JOHN H., Professor of Education, Northern Illinois University

LESLIE, LARRY L., Chairman and Associate Professor of Higher Education and Research Associate, Center for the Study of Higher Education, The Pennsylvania State University

MAXSON, ROBERT C., Division of Education, Auburn University at Montgomery

SISTRUNK, WALTER E., Professor of Education, Mississippi State University

THRASHER, JAMES M., Director, Institute for Educational Management, United States International University

WILLIAMS, JAMES O., Chairman, Division of Education, Auburn University at Montgomery

WOMACK, DARWIN, Assistant Superintendent and Director of Construction Services and School Planning, Atlanta City Schools

WORNER, WAYNE, Head, Division of Curriculum and Instruction, College of Education, Virginia Polytechnic Institute and State University

Contents

Preface ix

1. Purposes, Problems, and Promises of American Education 1
   Walter E. Sistrunk and Robert C. Maxson

2. Theory In Educational Administration: A Systems Approach 25
   Howard J. Demeke

3. An Administrative Organization for Implementation of Educational Purposes 60
   James M. Thrasher and Lyal E. Holder

4. The Administrative Team Concept 76
   Charles Atwell, Dewey A. Adams, and Wayne M. Worner

5. The Administrator and the Changing School Curriculum 107
   Oscar T. Jarvis and Richard W. Burns

6. The Administrator and Instructional Innovations and Technology 138
   Larry L. Leslie

7. Managing the Instructional Team: Personnel Administration 172
   James O. Williams

8. Financing the Educational Program *Robert J. Garvue* 200

9. Serving the Client System *John H. Johansen* 238

10. Trends In School Plant Design: Facilities for Today and Tomorrow 256
    John W. Gilliland and Darwin Womack

Index 283

Preface

This book is designed to be a major guide and resource for all those who study educational administration. It is a collection of essays developed by writers who were chosen because of their expertise in specific phases of administration. The essays were written specifically for this book. Because of the quality of the participants, the editors believe this book will make a real contribution to the field of educational administration and may become one of the more scholarly works in this area.

Chapter 1 is an examination of the public school as a system within the broader context of the supra-system, with special emphasis on the problems and promises engaged in fulfilling the mission of American education. An administrative theory based on the general systems concept, including a framework for decision making, communications, compliance, and power distribution and utilization is developed in Chapter 2. Analyzed in Chapter 3 is the design of administrative organization as a means of accomplishing the purposes and promises of American education while solving some of its problems. Chapter 4 proposes a rationale for an administrative team approach to achieving unity of purpose, while retaining unity of command and providing for proper division of labor, fixing of responsibility, and delegation of authority. Chapter 5 takes a look at the school curriculum with emphasis on trends for the future and a realistic means of administering them. Chapter 6 is a survey of innovative teaching techniques and supporting technology with emphasis on how they affect the administrative process. A rationale for administrative leadership as a basis for personnel management, including an analysis of relevant issues such as teacher evaluation, accountability, professionalism, teacher strikes, and negotiations are outlined in Chapter 7. Chapter 8 is an inspection of the problems involved in financing the educational program, with emphasis on present trends and how they are impacting upon the school administrator. Chapter 9 is a study of the role the educational leader plays in bridging the gap between the school and the community to better meet the needs of both systems. An examination of innovative trends in educational facilities and maintenance and the role of the administration in basing these types of decisions on the needs of the instructional program is made in Chapter 10.

This book is our response to a felt need for a scholarly textbook based on a systems approach to educational administration. We owe a particular debt of gratitude to the contributors who took time from extremely demanding schedules to assist in the writing of this book. We are thankful for the work of our secretaries: Bonnie Blackstock and Delaine Gardner, and to our respective universities, which made the work possible. A special debt is owed Dr. James O. Williams, an administrator's administrator, for providing valuable guidance and advice throughout the entire project.

Auburn University at Montgomery   ROBERT C. MAXSON
Mississippi State University   WALTER E. SISTRUNK

WALTER E. SISTRUNK
ROBERT C. MAXSON

Purposes, Problems, and Promises of American Education

"If a Nation expects to be ignorant
 and free, in a state of civilization,
 it expects what never was and never will be."
                    Thomas Jefferson

INTRODUCTION

Jefferson wrote his friend George Wythe, "Preach, my dear Sir, a crusade against ignorance; establish and improve the law for educating the common people."[1] He foresaw the heavy tax burden which would become necessary to free the minds and spirits of all the children of all the people; but he believed the cost of education would be less than a fraction of the cost of the tyranny and chaos caused by ignorance.

Few people have ever put more reliance on education as a means of achieving national security and greatness than have the people of America. Even fewer peoples have viewed education (formal schooling) as a necessity to achieving national greatness. The American dream always was that "my child can be better than me," and this is still our national and individual dream. Education in the United States was intended to be the ladder to the stars! Indeed, it has been for many. Many others, especially the children of the poor and of darker skinned ethnic minority groups, have found free, public, universal education in the United States to be little more than a dead end street. Far from finding schooling to be a means of

---

1. Saul K. Padover, *The Genius of America* (New York: McGraw-Hill Book Co., 1960), p. 60.

rising socially and economically many people have found that schooling only taught them enough to understand their frustrations and blighted hopes.

It is in the nature of any free society for the citizens to expect much of education and schooling. No people ever dreamed greater dreams than common Americans, and none have ever been able to use schooling more effectively as a means to realizing their dreams. We expect much of our schools and are all too quick to blame them when things seem awry.

A main purpose of all education in a free society must be to teach students how to govern themselves; hence effective schooling in our society should give some attention to teaching by precept, example, and exposition how to live and remain free as they grow toward their several individual potentials in a free society.[2] The United States is a pluralistic society with social, cultural, political and ethnic roots in many lands throughout the world. Individuals, institutions, and groups with divergent views, beliefs, and needs exist within our society in response to felt human needs and problems. These felt needs and problems create conflict over the purpose of education, about the control of education, and about the means of education. Differences over the purpose and about the governance of education are often alarming; but such differences and conflicts are natural in a pluralistic society. These differences need not be alarming; they can serve as a medium for the exploration and trial of new ideas, new approaches to schooling, and to development of more meaningful purposes. Developing more acceptable and meaningful purposes for schooling is essential to maintenance of the vitality and strength of a dynamic, democratic, and viable society. Consensus arrived at without prior difference and conflict is of dubious value in a free society.

The main struggle going on about schools today is centered on what Luvern Cunningham calls governance.[3] According to Cunningham, who is to control the schools is the central issue troubling the nation and the educational community.[4] Among the unique features of the organization and administration of schools in the United States is its decentralization and its control by the *people* rather than by the government or some bureau of government.[5] Cunningham asserted that the major struggle over education in the United States, at least as it affects administrators, is over

---

2. Walter E. Sistrunk and Robert C. Maxson, *A Practical Approach to Secondary Social Studies* (Dubuque, Iowa: Wm. C. Brown Company Publishers, 1972), p. 19.

3. Luvern Cunningham, *Governing Schools: New Approaches to Old Issues* (Columbus, Ohio: Charles E. Merrill Pub. Co., 1971).

4. *Ibid.*, pp. 20-21.

5. Edgar L. Morphet, R. L. Johns, and Theodore Reller, *Educational Organization and Administration* (Englewood Cliffs, N. J.: Prentice-Hall, Inc., 1967), p. 17.

who shall control the schools, how they shall be controlled, and for what purpose shall they exist. Certainly Cunningham has developed an important point, and while governance may not be the only issue with which we administrators must deal, it is, nevertheless, one of the major ones. It is impossible to perceive who will ultimately control American public schooling in the decades ahead and it is equally risky to predict the purposes which the controllers will see as vital. Efforts to predict the shape of education in the future or even its promises may fall far wide of the mark. However, the need to discern the shape, control, and purposes of public education in the future is great.

Administrators need better tools for futuristic educational planning. We need political sagacity, social acumen, and sensitivity to the needs of widely different and violently competing groups if we are to maintain control of the course of education. General systems theory, systems analysis and specific techniques of analysis such as PERT, PPBS, and CPM offer administrators more effective research and technical tools for forecasting future needs, trends and problems! It is imperative that administrators learn more modern means of delegating responsibility, dividing labor, involving people, collecting and analyzing data, and more effective means of advising school boards, legislatures, the congress and various agencies of the government.

Effective education for citizenship, and vocational competency represents the difference between progress and chaos. Disaster awaits any nation that leaves the schooling of its young to chance. Education in the United States long ago set out to release the potential of each student so that he may grow more unique each day. This purpose holds out enormous promise to those who are able to perceive meaning in schooling as offered.

THE STATE OF EDUCATION

The promises of education in America remain unfulfilled for many, as indicated in the following excerpts from a timely report by Dr. Sidney P. Marland.

THE CONDITION OF EDUCATION IN THE NATION

"The time seems propitious for a report on the condition of education in the Nation. The long swell of history appears at this moment to have lifted us above the turbulence of recent years and positioned us to appraise with some reasonableness the present condition of the educational enterprise. It is a commanding view, a prospect at once gladdening and disturbing.

We can take legitimate satisfaction from the tremendous progress of recent years. The sheer size of the American commitment to education is amazing, with over 62 million Americans—More than 30 percent of the population—actively engaged as students or teachers. More than three million young men and women graduated from high schools throughout the country in 1971, as contrasted with fewer than two million 10 years ago. Nearly 8.5 million students are enrolled in higher education as contrasted with slightly more than four million 10 years ago. Size apart, our educational enterprise is also far more nearly equalized, with academic opportunity extended for the first time in our history to large numbers of black, brown, and Spanish-speaking people. Total black enrollment in colleges and universities, for example, has more than doubled since the mid-60's to nearly 500,000 today, though much remains to be done for the advancement of our minority young people before we can rest.

We can be proud of the willingness and rapidity with which education has begun to move to meet the extensive and unprecedented demands being made upon it. . . .

But, viewed objectively, the great flaws of the educational system, the great voids in its capacity to satisfy the pressing requirements of our people press us to set aside our pleasant contemplation of our successes. Sadly, the quality of education a person receives in this country is still largely determined by his ability to pay for it one way or another. As a consequence, "free public education" has a connotation in, say, Shaker Heights far different from what it has in the city of Cleveland, and a boy or girl from a family earning $15,000 a year is almost five times more likely to attend college than the son or daughter in a household of less than $3,000 annual income.

We know that ours is the greatest educational system ever devised by man. But it falls short of our aspirations. We must improve it.

Like our system of representative government, the American education system is too vital for us to ignore or abandon because it has faults. It is time to set about, in an orderly fashion, making the system work better so that it will accomplish what we want from it.

American education has undergone over the past 10 years probably the most wrenching shakeup in its history. Education has been charged with inefficiency, unresponsiveness, and aloofness from the great issues of our society, perhaps even lack of interest in these items. These charges, in some instances, have undoubtedly been true. But in most cases, I insist, the schools and those who lead them and those who teach in them are deeply, painfully, and inescapably concerned with the great social issues of our time and the part that the schools must play in resolving them.

The depth of the schools' contemporary involvement becomes strikingly apparent when it is compared with the false serenity of education as recently as 15 years ago, when it was in the very absence of stridency and criticism that our real problems lay. Public discontent with the education of the 1970's was bred in the synthetic calm of the 1950's and before.

This movement from serenity to discontent, from complacent inadequacy to the desire for vigorous reform has not been accomplished easily. Some reform efforts, conceived in an atmosphere of hysteria, have failed while others have succeeded splendidly. But after many stops and starts, false expectations and disheartening letdowns, we have arrived at a time and place in which, I judge, educational reform at all levels is the intent of all responsible educators. As a consequence, truly equal educational opportunity for all young Americans is now a feasible goal.

We are going through a period of intensive concern with the poor and the disadvantaged. Since 1965 under one program alone, title I of the Elementary and Secondary Education Act, the Federal Government has invested more than $7 billion in the education of children from low-income families. A number of states have made significant companion efforts. Admittedly, our success in increasing the academic achievement of the disadvantaged child has been marginal. But prospects for future success are increasing because the education profession itself, at first prodded into this work by such outside forces as the drive for civil rights, is now substantially dedicated to the redress of educational inequality wherever it may be found. This is a dramatic turnaround from the early and mid-60's when we tolerated the fact that certain of our citizens were not profiting to any measurable extent from the schools' conventional offerings and when we were content to permit these citizens to become the responsibility of unemployment offices, unskilled labor pools, and prisons. This time has passed, and we now accept the proposition that no longer does the young person fail in school. When human beings in our charge fall short of their capacities to grow to useful adulthood, we fail.

Rough events of the past decade, then, have brought the educators of this Nation to a beginning appreciation of just what thoroughgoing education reform really means. A giant institution comprising 60 million students, 2.5 million teachers, and thousands of administrative leaders cannot remake itself simply because it is asked or even told to do so. Tradition has enormous inertia, and wrong practice can be as deeply rooted as effective practice. The past decade, in sum, has been a time of trial and error, a time in which we have plowed and harrowed our fields. Now we must plant deeply to produce the strong roots of a new American education.

As we look to the future, we are able to state with far greater clarity the reasons why we are educating our citizens than we could 10 or 20 years ago. We are educating a total population of young people in the elementary and secondary schools and we are no longer satisfied that 30, 40, or 50 percent of it should not really expect to complete high school. And if we are educating for the fulfillment of all the people of our land, we certainly cannot halt at the secondary level, or even the level of higher education, but must look to the arrangements for continuing adult education over the years. Increasingly, we are persuaded as a Nation that education is not reserved for youth but is properly a lifelong concern. . . .

We must be concerned with the provision of exciting and rewarding and meaningful experiences for children, both in and out of the formal environment of classrooms. When we use the word "meaningful," we imply a strong obligation that our young people complete the first 12 grades in such a fashion that they are ready either to enter into some form of higher education or to proceed immediately into satisfying and appropriate employment. Further, we now hold that the option should be open to most young people to choose either route.

We must eliminate anything in our curriculum that is unresponsive to either of these goals, particularly the high school anachronism called "the general curriculum," a false compromise between college preparatory curriculum and realistic career development. If our young people are indeed disenchanted with school —and more than 700,000 drop out every year—I suspect that it is because they are unable to perceive any light at the end of the school corridor. They cannot see any useful, necessary, and rewarding future that can be assured by continued attendance in class. The reform to which we must address ourselves begins with the assurance of meaningful learning and growth for all young people, particularly at the junior high and high school levels. Students frequently ask us why they should learn this or that. We who schedule these courses and we who teach them should ask ourselves the same questions and have the wisdom and skill and sensitivity to produce good answers.

Courses of instruction, books, materials, and the educational environment —all should relate to the student's needs, answering some requirement of his present or future growth, irrespective of custom or tradition. We as teachers in today's educational setting cannot win the response of our young people by perpetuating formalized irrelevance in classrooms. Seemingly irrelevant expectations must be made relevant by the teacher. This is the nature of teaching.

We are obliged not simply to provide education but to provide very good education. The success of our efforts to find ways to teach more effectively will depend upon the quality and application of our educational research, a pursuit that has absorbed more than $700 million in Office of Education funds over the past decade and will, I am determined, take an increasing share of our budget. We need to know how we can develop the child of deep poverty, the minority child, the child who has been held in economic or ethnic isolation for generations, the child without aspirations in his family or in his environment, the child who comes to school hungry and leaves hungrier. We must discover how to develop the five million American children who bring different languages and different cultures to their schools. They need special help. Nor can we ignore the gifted child, possessed of talents that we know frequently transcend the ability of his teacher. . . .

I believe the Federal role in education should be one of increasing the effectiveness of the human and financial resources of our schools, colleges, and universities. The present overall level of Federal assistance to education is something less than seven percent of our total investment. I envision the Federal share rising eventually to three or four times that level. But first the Federal Government must conduct centralized research into the learning process and deliver the results of that research convincingly and supportively to the educational institutions. We are constructing a nation-wide educational communications network to disseminate proven new practice in order to move the art of education from its present condition to one of the increased quality that we demand of ourselves. We must proceed more swiftly to implement the products of research without stopping to redefine every goal and every process at every crossroad in the country.

The Federal role calls for greatly increased technical assistance to States and local school systems to insure the delivery of new and better ways to teach and learn. As conductor and purveyor of educational research the U. S. Office of Education will, I hope, earn the faith and trust of the States and communities so that newly researched and validated program models stamped "O. E." will be swiftly and confidently put to use in our cities and towns, creating the overall climate of change that we ask.

Most of all, I ask that the Office of Education provide national leadership. Services, yes; supporting funds, yes. But I hold that this Office, made up of nearly 3,000 people, must have a larger and more effective role. If our situation changes over the next year or two as I hope it will, and we are able to diminish substantially our preoccupation with administration and paperwork, hundreds of Office of Education staff members will be freed to bring leadership, technical assistance and stimulation to the States and localities. . . .

The United States of America will celebrate its 200th birthday in 1976. I would suggest this bicentennial year as a useful deadline against which we can measure our capacity to effect change and our sincerity in seeking it. The years remaining before the bicentennial constitute a relatively brief time in the history of the American educational enterprise. Yet it is a particularly crucial time in which, I am persuaded, we can accomplish as much as—and more than—we have managed to achieve in the past 20 years, or perhaps the past 100. My reason for optimism resides in my belief that, big as this Nation is, it is ready for change.

Our search for the education of 1976 is well begun. We know it will be innovative and efficient, yet characterized by good schoolteacher common sense. We know it will be flexible, responsive, and humane, that it will serve all the children of America, preparing them to meet universal standards of excellence, yet treating each in a very individualized and personalized way. We know that in 1976 our system of education will be considerate of the differences among us, adaptable to our changing expectations, and clearly available and clearly useful to all who seek it.

More than ever before, the substance of America's future resides in our teachers. The enormous success of our system of schooling in the past 195 years has brought our Nation to a pinnacle place among nations. The next five years should be viewed as the time in which the educational successes and satisfactions that have enlightened and undergirded the lives of the great majority of our people must now be extended to enlighten and undergird the lives of all. More than ever, this is the time of the humane teacher."[6]

6. Sidney P. Marland, "The Condition of Education in the Nation," *American Education*, Vol. 7, No. 3 (April 1971), pp. 3-5.

Commissioner Marland made a strong case for the worth of our schools without claiming that they are perfect. He pointed out the magnitude and the complexity of the tasks we face in the decades ahead if we are to fulfill the promises of the American dream for the children of all the people. Marland delineated some proposed courses of action which he believed deserved careful consideration. The solutions he proposed are specific, concrete and thus evoke hopes, fears, agreement, and disagreement.

A more general view of possible solutions to our educational problems might involve use of general systems theory and systems analysis to examine appropriate roles and actions for the institutional system (voters), the managerial system (administration), the technical system (teachers, aides, etc.) and the captive client system (students). Administrators need to learn how to use general systems theory to understand the relationship of the school to its several communities. It is essential that systems analysis come into common usage as means of helping the managerial system make more effective use of the time and talents of those in the technical system in order to better serve the needs of the school's clients.

FREE PUBLIC SCHOOLS

Education has had a special mission to perform in America from the very beginning. Democracy cannot function without an enlightened electorate. A nation cannot be built without a common language, history, culture, and purpose. Immigrants by the millions had to be acculturated, farm people had to learn the skills needed to work in industry. European class distinctions would have eroded the infant democracy if effective, free schooling had not been open to all people. No other nation ever demanded or expected so much of its schools as did the United States. No nation was ever better served by its schools or its educators.[7] The infant republic delegated to its schools the task of teaching democracy, nationalism, and equality; in short the purpose of schooling became the making of effective *free citizens*.

Few realize that only three universities of the mother country are older than Harvard. Even fewer are aware that by 1776 there were more colleges in the colonies than in Britain.[8] Commager asserted that America put its faith, from the first, in education and he concluded that our faith

7. Henry Steele Commager, "Free Public Schools—A Key To National Unity," in *Crucial Issues In Education,* ed. Henry Ehlers (New York: Holt, Rinehart & Winston, Inc., 1969), pp. 6-7.
8. *Ibid.*

can be justified on purely scholarly grounds. Imagine what life in the United States, and indeed the world, would be like without our great land-grant colleges and state universities with the free, public, compulsory schools which feed them. No nation ever invested so much in education, no experiment ever returned so much for the capital invested! Even a casual examination of the G.I. Bill and the contribution of its beneficiaries to the tax structure of the United States makes one keenly aware of the pervasive influence of education on our place in the world.

Several tasks were thrust upon the schools of the nation, and some remain uncompleted. Let us examine a few of them. The earliest task we set for the schools was to *provide an enlightened citizenry* as a prerequisite to functioning democracy. The second task was the creation of national unity. *E. Pluribus Unum*, out of many, one, was our motto and the goal of our schools.

About 1840 America began to be flooded with a new type of immigration; people who spoke other languages and had other religious backgrounds became the rule rather than the exception in the United States. Once again we called on the schools to Americanize the immigrants. No other nation has ever absorbed so many and so varied a people so rapidly. The fourth task assigned the schools was overcoming a divisive tendency growing in the young nation. These divisive forces in the young heterogeneous society did not prevail.[9] The United States persisted and yet other tasks such as vocational competence, racial desegregation, technical proficiency, and cultural tolerance have been assigned to our schools.

Goodlad asserted that the most important task for our schools in the next few years is the daily practice of the qualities of compassion, sound judgment, flexibility, humility, humane outlook and self-renewal. He advocated the infusion of values as means of education previously thought of as ends.[10] Free societies cannot afford to subvert the means of freeing individuals to any ends no matter how desirable a particular end may be.

Survival of democracy depends upon the people maintaining the ultimate power of decison making. There are those who assert that most decisions about schools are made by the courts. Others assert that various pressure groups have usurped the board's powers of decision making. Some teacher organizations are asserting that their members, and only their members should have the power to make decisions regarding the curriculum. Such a contention is antithetical to the basic American prop-

9. *Ibid.*
10. John I. Goodlad, "The Educational Program To 1880 and Beyond," *Implications For Education of Prospective Changes In Society,* ed. Edgar L. Morphet and Charles O. Ryan (New York: Citation Press, 1967), p. 47.

osition. Teachers and administrators should offer professional advice to school boards and legislators; but no group should usurp the power of the voter to control educational policies through his duly chosen representatives.

The American proposition is that *the people,* not the *professionals* make policy. It was a new proposition when first advanced, and it is still on trial. Most nations still refuse to accept the view that man can or should govern himself. Whoever controls education controls the future. We cannot afford to delegate such an awesome responsibility to some pressure group, or to the profession. Educational decision making is a public trust held by all citizens and delegated by them to legitimate legislative and executive officials. We must not permit this public trust to pass to any group by delegation or by default. Use of adequate systems techniques will help provide the citizenry with sufficient data from which sound educational decisions can be reached.

It seems reasonable to assert that the purposes of education asserted in the Seven Cardinal Principles, Ten Imperative Needs Of Youth, The President's Commission on National Goals, and The Central Purpose of American Education, are as valid today as ever. Many groups such as the Educational Policies Commission, ASCD, NASSP, and AASA have formulated goals or purposes of American education. Many individuals have also set forth purposes and goals for education. All these sets of goals have common elements, all are valid for today's world; but few, if any, are comprehensive or sufficient.

Schools in the United States have been sensitive to the needs and opportunities of a changing society from the beginning. Two primary purposes were established in the debate regarding public education in the early history of this country. The individual was to have the opportunity to get an education which would equip him to develop his potential and become a happy and productive citizen. Another goal of education was to make literate citizens of the young. A democracy can only exist where there is an educated citizenry.[11]

The idea of democracy is not simple and the practice of democracy is not easy. There must be responsibility with freedom for without it there is no freedom. The rights of others must be taken into consideration as well as the rights of the individual. It might be said that the individual must be taught how to acquire the skills of critical thinking in order to develop responsibility. These values must be taught in the school and the

---

11. Ralph W. Tyler, "Purposes for Our Schools," *NASSP Bulletin—Looking Ahead,* Vol. 52, No. 332 (Dec. 1968), p. 11.

home along with the facts and skills that parents and teachers feel important.[12]

Education has been the cornerstone of the development of freedom and progress in the United States. Early in the history of our country our forefathers realized the importance of establishing schools to educate their young people. Side by side the history of education and the history of our country have grown and have changed. The doors of the schools have afforded the citizens of our nation an opportunity to seek knowledge, to develop skills and training, to have a better life, and to become better citizens. Increasing the quality and availability of education is vital to our national security, our domestic well being, our economic growth, and our position as a world power.

At the same time we have continued to be a nation that has tolerated ignorance, illiteracy, delinquency, unemployment, school dropouts, and dependency upon welfare benefits. America is a unique civilization, not quite like any other in the world. No other nation has the technology, power, and freedom to do so much for so many; and yet, there are so many persistent inadequacies in our educational system because we expect so much more from it than other countries expect of theirs.

The shape of American schools in the future will be determined by national purposes for education. Education is shaped by present problems and expectations of future problems. There are many unfulfilled promises in America, education can help us deliver on some of these promises.

PURPOSES OF EDUCATION

"Prior to recent times the definition of educational purpose was stated in operational terms. Education in ancient Israel was to teach the law, education in Sparta was to train soldiers, education after Christ was to impart Christian values."[13] America has always been a nation of seekers for freedom from every culture in the world. The task of the educational leaders of our country has been to take all the children of these immigrants and weld them together to make Americans, yet, encourage them to retain many of their customs and traditions from the old country. The acculturation of the children of immigrant groups became one of the first purposes of our educational system.

12. Jane Franseth, *Supervision as Leadership* (Evanston, Illinois: Row, Peterson, and Co., 1961), pp. 10-11.
13. John A. Bartky, "An Operational Definition of Public School Purpose," in *Views of American Schooling*, ed. Jonathan C. McLendon and Laurence D. Haskew (Glenview: Scott, Foresman and Company, 1964), p. 62.

No other nation has staked so much of its hopes upon the education of its people as has the United States.

Our early political leaders favored a system of popular education to assure an adequate supply of enlightened citizens who would vote wisely, to insure a stable government, and to preserve democracy. These leaders saw very clearly that only educated men should be able to judge political issues with discrimination; choose sound doctrines and reject false ones; restrain the human urge to get something for nothing from the government; and choose honest and able candidates for public office, rejecting the demagogue and the cheat. This was a radical departure from the existing concept of the nature of man, particularly from the existing concept of the capacities of the people. The prevailing doctrine, at that time, was that only the upper class—privileged and the property owners—were fit to govern; that the common people were without capacity for leadership.[14]

Thomas Jefferson perceived that man's failure to reach his full potential was not due to his inferior nature, but rather to his lack of education. The fact that the United States is the oldest republic in the world has a connection with the fact that we also have the oldest universal public school system.

The pattern of American schools is determined by the commitments our society has. Fundamentally, education exists in the United States to develop citizens who will be able to participate effectively in a democratic society and make decisions that will contribute to a realization of the democratic ideal. Two basic elements of this ideal are that the state exists to promote the welfare of the individual and that each person must be free to choose the contribution he will make.[15]

Today education for all is not the only purpose of our school system. The world of learning is no longer a neat, compact, cozy place. Modern technology has made the whole world available as a classroom. Educators have begun to realize that quality is more important than quantity. The individual potential of each child needs to be discovered and developed to enable him to take on a sense of identity and accept the values that will make him a worthwhile adult.

It is time for us to realize that the purposes and goals we had in the past will not suffice for the present. Our educational system has had to seek new ideals for the future in order to survive and continue to meet the

14. T. M. Stinnett and Laurence D. Haskew, "Education and the Promise of America," in *Views on American Schooling*, ed. Jonathan C. McLendon and Laurence D. Haskew (Glenview: Scott, Foresman and Company, 1964), p. 151.

15. Kimball Wiles and Franklin Patterson, *The High School We Need* (Washington, D.C.: Association for Supervision and Curriculum Development, 1959), p. 1.

needs of the nation. Educators must wear two hats, "keep everyone up-to-date in order to cope with the personal problems of the age and culture, and challenge the specialized talents of the individual."[16] Competence of the person *after* education is the ultimate test of any educational system or experience. We must prepare students to be competent in a world vastly different to any ever before known. Students must be prepared to be competent humans, vocationally competent, and competent citizens.

The problem of defining purposes and goal setting may be the most difficult problem of the seventies and the decades beyond.[17] One nearly insoluble problem is how to finance the schools, another is how to manage the personnel who work in them. Finance and disagreement over who shall determine purposes and goals (thus struggle over what the goals will be) may well determine the things we do, the way they are done and the goals we seek. Educational policy and decision makers face monumental and virtually insoluble tasks in the years ahead. The sheer quantity of data we must deal with dictates the need for more effective school management and leadership. Use of the tools provided by general systems theories and systems analysis coupled with the ready availability of data processing may be the only way harassed administrators can provide dynamic equilibrium to the shifting school scene.

CONTROL OF EDUCATION

The American school system has served us well throughout our history largely because of the blending of the policy making function in the hands of laymen with policy implementation by professionally trained administrators.[18] The United States constitution has left the government of schools to the states by inference. There is no mention of education in the Constitution, thus the states assumed plenary power because of the tenth amendment. There are some common features of educational control in most states such as the actual control of education by local boards. Only Hawaii has no local school districts; however, Alaskan state government controls all schools in the state. In contrast to such centralization, some mid-western states have several thousand local school boards. The role of the Federal government is in rapid transition from advisor, to persuader, to stimulator, to controller of some school functions.

16. J. Lloyd Trump and Dorsey Baynhan, "The School of Tomorrow," in *The Teacher and the Taught*, ed. Ronald Gross (New York: Dell Publishing Company, 1963), p. 285.
17. Cunningham, *Governing Schools*, p. iv.
18. *Ibid.*, pp. 201-202.

Public schools in the United States are part of the society at large, thus they must ultimately reflect the strains and the stresses of that society. Most public school systems have been experiencing great stress during the last decade. These stresses stem from the changes in the legal structures provided for governing schools. Some call these shifts of power and control from the Courthouse to the Statehouse to the Nation's Capitol *federalism in transition*. Many local boards have been unresponsive to the needs and desires of their clients because of insufficient size, or money, or because they did not recognize needs. Whatever the cause of local unresponsiveness to clients desires and needs, the clients have turned to the states for help and when they found the states unresponsive they turned to the national government as an instrument which could meet their needs.

School districts have been growing in size and complexity for years and the trend continues. Many districts are now also faced with the need to decentralize, thus there are stresses because of centralization and decentralization. Many private groups are engaged in efforts to exert more influence on local school boards.

The United States has become a society where it is increasingly difficult for an individual to have much impact on government or on schools. This coupled with the breakdown in a sense of primary communities has led many citizens to become active in various groups designed to influence the local school boards. Often a person may be an active member of several groups whose goals are in active conflict. Such membership causes strains in both the personal and public life of the individual. These strains can cause people to behave irrationally and can cause the demands of the pressure groups to which they belong to act selfishly, irrationally and even illegally. School administrators are subject to a multitude of such pressures, stresses and strains on a daily basis. It has become commonplace when an administrator meets an old friend at a conference and asks, "how are things?" the reply will be, "quiet when I left."

The profession of teaching is reaching for maturity which has led teacher organizations to demand more control over what they teach and how it is to be taught.[19] Teachers are demanding autonomy and in some cases outright control of educational decision making. Teacher demands for power on the one hand and the public outcry for accountability on the other hand are creating great stress in many school systems.

The public has evidenced an increasing disenchantment with public schools during the past decade. Few school bond issues are passing, defeat of incumbent board members and elected school superintendents has become common place. The public has shown interest in performance con-

---

19. *Ibid.*, p. 207.

tracting, the voucher plan, private schooling, teacher evaluation and almost any other proposal that promises predictable results. There are hundreds of organized citizens groups in the nation protesting one or another aspect of the public schools. There has been a growing isolation of the school system from the client system which legitimatizes it. Many school people are not only isolated but also insulated from substantial segments of the population at large. These stresses have caused some administrators to become unduly sensitive to any information received from organized public groups. Such paranoia has caused a growing indifference to the educational welfare of students as individuals and an unwarranted sensitivity to students as group members.

Student unrest, disenchantment, and even crime, in school situations has become common place. In many urban schools criminal activities are intruding into school activities, sometimes even into classroom teaching situations. There is substantial student resistance to study and learning as well as to being governed and controlled. Students are disenchanted with school and are very vocal about it.

Science, technology, and instructional innovations introduced into the schools to help solve some of the problems and unrest described above have often been proved to be more bane than boon. Students often view gadgetry as a further depersonalization of education. Some of the innovations have created far more problems than they have solved. It was long thought that per pupil expenditure was an adequate gauge of educational excellence. It has turned out that money does not always improve education, just as it has turned out that larger schools are not always better ones.

The nation suffers from a sense of personal and national malaise. Why should public education, since it is a sub-system of the country, not also suffer some sense of unease? The sovereign American remedy for social ills has always been *to pass a law*. We are hastily passing many state and federal laws designed to ameliorate educational problems. Local boards and school officials scurry frantically about with stop gap measures designed to make the sense of unease go away. However, no law, no policy, no goal will do much toward resolution of educational problems so long as the source—society—remains filled with a sense of malaise. Administrators need to understand that states and local districts are part of the larger social system thus subject to the same ills and problems.

THE ISSUES

Although the issues are many and varied, it appears that most of them can be subsumed under the following major headings: control or governance, the roles of individual citizens and groups of private citizens,

16 AMERICAN EDUCATION

centralization, subdistricting or decentralization of very large school districts, the purposes of schooling, administrative and board roles, finance, and legitimation of board, administrator, and teacher roles.[20] Cunningham made a number of proposals about defining the roles of these individuals and groups.

The *Eight State Project* found twelve significant changes going on in society and asserted an implication for school administration from each.[21] These changes and their educational implications are asserted below.

1. Twenty-five per cent population increase in thirteen years.
   Implication: It will be difficult to provide adequate urban services.
2. The growth of knowledge will continue.
   Implication: Improved storage and retrieval systems will be needed. Books won't be enough.
3. Instant communication in most homes.
   Implication: Educators will need to use such devices as part of teaching.
4. The rapidity of transportation will increase greatly.
   Implication: Geographic mobility of population, already apparent, will increase thus increasing demands for a national curriculum.
5. The gross national product may grow as much as fifty per cent.
   Implication: The proportion awarded education should be fixed.
6. The work force will increase about one third.
   Implication: Continuing education and re-training will grow in importance.
7. Conflicts and ideology will increase. Differences between those who advocate change and those who are more conservative will be greater.
   Implication: Concepts to be taught will have to be more carefully selected.
8. Individual productivity will increase by about one third.
   Implication: Education will become an end as well as a means.
9. Men's biology will remain about constant but their environment will change greatly and rapidly.
   Implication: The gap between parent and child will grow greater and education must become a means of closing it.

20. *Ibid.*, pp. 20-23.
21. Paul A. Miller, "Major Implications for Education," in *Implications For Education of Prospective Changes in Society*, ed. Edgar L. Morphet and Charles O. Ryan (New York: Citation Press, 1967), pp. 20-22.

10. Governments of all kinds will impinge much more often and closely on individuals yet will seem more impersonal and farther away.
Implication: Natural and status leadership must not lose touch with the great mass of people. Education must teach people how to govern and how to be governed.

11. Nongovernmental organizations will play a more important role in purveying ideas and maintaining governmental balance.
Implication: Students must be taught the importance of organizations.

12. There may continue to be negative problems, such as war, catastrophe, racism or depression.
Implication: Students will continue to need help in understanding the human condition and improving it where possible.

These twelve significant changes and their implications have been reported in the words of the authors rather than those of the Eight State Project. However, every effort was made to keep the sense of what the project reported. The Eight State Project went on to state a variety of implications for the Federal government, the several states, and for local school districts. These implications were not included here in the interest of conserving space but the reader can find them in *Implications For Education of Prospective Changes in Society* which is referenced at the end of this chapter.

It seems clear from a review of the literature that educators have a reasonably firm grasp of what the problems of education are; but the grasp of what to do about them is far less common or adequate. Until the conflict over who is going to control the goals of education is resolved, any predictions about the nature of future education is speculative at best.

Cunningham recommended a number of intriguing proposals in his book on school governance.[22] R. L. Johns in Chapter 14 of the report on the Eight State Project has made a comprehensive proposal for changes in legal responsibilities at the Federal, state, and local levels.[23]

ORGANIZATION AND RESPONSIBILITY FOR EDUCATION

Legal authority and responsibility for educational organization and control is vested by inference in the several states. State organization and

---

22. Cunningham, *Governing Schools*, pp. 142-199.
23. R. L. Johns, "State Organization and Responsibilities for Education," in *Implications For Education of Prospective Changes in Society*, ed. Edgar L. Morphet and Charles O. Ryan (New York: Citation Press, 1967), pp. 20-22.

control of education cannot be examined without consideration of the United States Constitution, Supreme Court decisions, United States laws, and the regulations of HEW, because the division of power between the national government and state governments has been severely eroded and dramatically altered during the years of *federalism in transition* which we have experienced since the depression of the 1930's. Similarly one cannot understand the states' control of education without consideration of local responsibilities and duties.[24] An adequate description of the organization and control of education in the United States would require many volumes since there is *no* American educational system in the formal, legal sense. Rather, education in the United States is composed of a system of thousands of infinitely varying sub-systems.

Even a cursory examination of the present legal basis for the organization and control of education reveals many problems and constraints that must be lived with or changed.[25] There are countless sub-systems within both the *managerial system* (superintendents, principals, supervisors, et al.) and within the institutional system (state and local boards, advisory committees, etc.). Each school is a sub-system of a larger unit, the local school system, which is in turn a sub-system of the state school system which is a sub-system of state government. Persons within a sub-system interact more freely and openly with each other than they do with persons from other sub-systems. Supra-systems are the environment of any sub-system. A sub-system survives and prospers only as it is able to understand its environment and adjust to it. Each sub-system strives to maintain equilibrium, a difficult task when the environment (supra-system) is out of equilibrium. No organization can survive if it cannot maintain a relatively stable equilibrium in the face of a changing environment. All surviving organisms and organizations have adapted to change and all that hope to survive must continue to adapt to change. Even the least complicated of living organisms adapt to changed environments, however, most such organisms are unable to control or manipulate their environments—they simply adapt or perish. Man, however, is a far more complex organism and his organizations and institutions are more complex. Man not only adapts, he also adapts his environment to his own uses. The same is true of a sub-system such as a school; it must adapt to its changed environments (such as government, community, society, or nation); but it is not content to simply adapt, a school will actively seek to alter or change its environment to meet its own needs. Thus in a very real sense man is the creator of his own environments, therefore, to some degree, he is his own creature.

24. *Ibid.*, pp. 245-248.
25. *Ibid.*

Schools or school systems are no different. They, like the people who compose their membership, seek to alter or create a more favorable environment, and are often quite successful at it. As a result schools are, to some degree, their own creatures and their own perpetuators, and like men they wish to be creative and they wish to live forever.

According to the general systems theory, all forms of animate or inanimate matter can be represented as systems which are interrelated. Social and/or political organizations can be understood in the same way. Every system is a sub-system of a larger, more inclusive system which makes up its immediate environment. Every system is composed of subsystems. Schools or school systems are social systems legitimatized and controlled by political systems. Schools or school systems are special purpose social and political sub-systems or units. A school is a system made up of a number of individuals and groups of individuals brought together by a supra-system to achieve a common purpose, though the members of the sub-system play widely varying roles and are assigned widely diverging tasks. An adaptation of Talcott Parsons general theory of administrative organization shows a school social system composed of:

1. the *client system* comprised of pupils or students
2. the *technical system* comprising teachers and other employees at the operational level
3. the *managerial system* composed of the superintendent, central office staff, principals, and other administrators
4. the *institutional system* comprised of the board, and the electorate.[26]

An older and simpler way of stating this theory is that those persons concerned with a school system comprise the voters and the board (policy makers), the administration (policy implementers), the faculty and staff (workers), and the students (learners). General systems theory helps us understand the school's environment, the interaction and communications networks within it, and helps us analyze problem situations. The real value of a systems approach to administration rests in its ability to help the administrator and his board gain a comprehensive, inclusive view of the organization's present condition, future problems, and some alternative solutions.

This book will devote most of its remaining space to illustrating ways of using the general systems theory and the tools of systems analysis as a means of helping the managerial system (administration) make more effective use of the goods and of the services of the technical system in the

26. *Ibid.*

service of the client systems needs, as determined by the institutional system. Every effort was made to help the reader more clearly discern the purposes, problems, and promises of education in the United States and some means of using systems theory to maintain dynamic equilibrium.

Johns proposed some means of achieving a more viable educational system, as did Cunningham. Commissioner Marland in the speech partially reproduced earlier in this chapter also pointed to some possible means of coping with educational demands. Some of Johns concepts are worthy of mention, though space precludes a thorough exploration of them. Johns proposes a role for HEW which he calls "creative Federalism." What he wants is a social system which is well staffed and a strong HEW capable of interacting with state and local school agencies.[27] Furthermore, Johns advocates a similar relationship between state departments of education and local boards. He perceives a crucial need for more effective provision of educational services from both the national and state levels. Cunningham recognized the problems of teacher militancy, client unrest, struggle of conflicting groups for control, weak local boards, need for centralization and decentralization, the impossible burdens of the superintendency, the weaknesses of preparation programs, and the most crucial problem of urban breakdown.[28] He made a number of specific proposals which he thought would alleviate some of the problems. Cunningham's most comprehensive and interesting proposal was a version of the Model Cities project.

It seems to the authors that the concept of an administrator as a sole proprietor, the person who does everything, is out of step with modern management theory. It seems curious that industry would have moved to the team management approach so long before education. It also seems curious that educational administration would remain the last great bastion of the "bottleneck" approach to administration. After all, Adam Smith recognized the need for specilaization and division of labor in 1776. General systems theories provide tools that will help us divide administrative labor more effectively, tell us more about the special competencies members of the team need, and will help us predict future needs based on changes within the system and its environment.

THE PROMISE OF GENERAL SYSTEMS

The scientific method was first developed about two centuries ago by combining logical analysis and natural science with the use of the laws

---

27. *Ibid.*
28. Cunningham, *Governing Schools*, pp. 221-231.

of probability.[29] The development of modern scientific procedures about one hundred years ago, as a part of the development of pragmatism, made possible the tremendous accumulation of knowledge and a far reaching technical revolution. These techniques have been influential on mathematics and science, as well. There have always been some observable similarities between components of theories in widely varying fields, but the development of theories of relativity, probability, evolution, and quantum measurement developed in recent years have enabled theorists to discover a well ordered set of principles which are applicable in many disciplines.[30] Granger calls this discovery metatheory. He views the application of systems theory to nuclear energy, rocket propulsion, data processing, medicine, and administration as the greatest discovery of the last twenty-five years.

PROMISE OF EDUCATION

We are begining to use and apply the concept of *systems* to educational administration through application of PPBS, PERT, and other systems analysis techniques. There are many other possible uses of general systems theory other than the making of budgets. Administrators need more specialized training in use of systems analysis as a means of making better predictions of future needs. A major change in the preparation, assignment and responsibility of administrative personnel seems needed.[31] The same unfulfilled promises of a free society are at work in schools as in other aspects of government. A systems approach seems to offer one of the better tools to help us meet our purposes and effectively fulfill the promise of America.

As remarkable as has been the contribution of public education to the welfare of the nation in the past, a much heavier load will be placed upon it in the future. Our schools must continue to enable the United States to lead the world in industrial and agricultural production by providing the scientific and the technological education which our society demands. We are told that within the next twenty years there will be 71 million more Americans than today with at least 70 per cent of them living in urban areas. The challenge of education is greater than ever. Will the schools be ready to meet this challenge?

29. Robert L. Granger, *Ed. Leadership: An Interdisciplinary Perspective* (Scranton, Penn: Intext Educational Pub. Co., 1971), p. 3.
30. *Ibid.*, pp. 3-5.
31. Claude W. Fawcett, "Ed. Pers. Policies and Practices in A Period of Transition," in *Implications For Education of Prospective Changes in Society*, ed. Edgar L. Morphet and Charles O. Ryan (New York: Citation Press, 1967), pp. 196-197.

Each generation of Americans paints a somewhat different picture of education within the framework of basic national needs and individual values. Horace Mann took into account the stirrings of the Industrial Revolution and the effects of the Jacksonian era. Dewey considered mass immigration and the spirit of the progressive era. Today educators need to take into account the new influences . . . . as they strive to move education forward in its dual role of service and leadership.[32]

One of the promises of education is that the focus will be on the individual. The curriculum must be differentiated to provide for individual differences if all are to have equal educational opportunities. Students will develop skills in critical thinking and problem solving to enable them to cope with the rapid growth of knowledge, the acceleration of change, and the complexity of choices that they must face.

Our schools have helped keep us free. In the future they will continue to be the great recreator of democracy. Education for all is the key to success and strength in a scientific and technological age far more than it was in the agrarian society of our forefathers. The good old days are gone; approached with intelligence and zest, the days of the future will be better. Such is the promise of American education.

Selected References

Alexander, William M., ed. *The High School of the Future: A Memorial to Kimball Wiles.* Columbus, Ohio: Charles E. Merrill Publishing Company, n.d.

Allen, James E., Jr. "Competence for All as the Goal for Secondary Education." *NASSP Bulletin—What's Right With American Education* 54 (May 1970): 9-17.

Austin, David B. "Thoughts and Predictions on the Principalship." *NASSP Bulletin—Looking Ahead* 52 (Dec. 1968): 141-150.

Austin, David B.; French, Will; and Hull, J. Dan. *American High School Administration.* New York: Holt, Rinehart and Winston, Inc., 1962.

Bain, Helen. "Self-Governance Must Come First, Then Accountability." *The Phi Delta Kappan,* Vol. 51, No. 8 (April 1970), p. 413.

Bennis, Warren G. *Changing Organizations: Essays on the Development and Evolution of Human Organization.* New York: McGraw-Hill Book Company, 1966.

Blue, Faye. "Curriculum Changes Necessary to Meet the Needs of Tomorrow." *The Delta Kappa Gamma Bulletin,* Spring 1970, pp. 6-9.

Castetter, William B. *Administering the School Personnel Program.* New York: The Macmillan Company, 1962.

Coleman, John E. "Origins of American Schools." *The Educational Forum,* May 1970, pp. 519-525.

---

32. Richard I. Miller, *Education in a Changing Society* (Washington, D. C.: National Education Association of the United States, 1963), p. 131.

AMERICAN EDUCATION

Combs, Arthur W., and Snygg, Donald. *Individual Behavior: A Perceptual Approach to Behavior.* New York: Harper & Row, 1959.

Counts, George S. "Where Are We." *The Educational Forum,* May 1966.

Cremin, Lawrence A. *The Genius of American Education.* New York: Random House, 1965.

Cunningham, Luvern L. *Governing Schools: New Approaches to Old Issues.* Columbus, Ohio: Charles E. Merrill Publishing Company, 1971.

Ehlers, Henry J., ed. *Crucial Issues in Education.* 4th ed. New York: Holt, Rinehart and Winston, Inc., 1969.

Estes, Nolan. "The Concept of Shared Power." *NASSP Bulletin—Common-Sense Priorities for the Seventies,* Vol. 55, No. 355 (May 1971), pp. 69-75.

Eurich, Alvin. "High School, 1980." *The Bulletin of the National Association of Secondary Principals,* No. 355 (May 1971), pp. 42-53.

Franseth, Jane. *Supervision As Leadership.* Evanston, Ill.: Row, Peterson and Co., 1961.

Gilchrist, Robert S. and Snygg, Donald. *New Curriculum Developments.* Washington, D.C.: Association for Supervision and Curriculum Development, 1965.

Goodlad, John I. "What Educational Decisions By Whom?" *The Education Digest,* Vol. 37, No. 2 (October 1971).

Granger, Robert L. *Educational Leadership: An Interdisciplinary Perspective.* Scranton, Penn.: Intext Ed. Pub. Co., 1971.

Grieder, Calvin; Pierce, Truman M.; and Rosenstengel, William Everett. *Public School Administration.* New York: The Ronald Press Company, 1961.

Griffiths, Daniel E., ed. "Behavioral Science and Educational Administration." *Sixty-Third Yearbook of the National Society for the Study of Education, 1964.* Chicago: University of Chicago Press, 1964.

Homans, George C. *The Human Group.* New York: Harcourt, Brace & World, Inc., 1950.

Koerner, James D. *Who Controls American Education?* Boston: Beacon Press, 1968.

Lieberman, Myron. *The Future of Public Education.* Phoenix Books. Chicago: University of Chicago Press, 1960.

Lucio, William H., and McNeil, John D. *Supervision: A Synthesis of Thought and Action.* 2d ed. New York: McGraw-Hill Book Company, 1969.

Marland, Sidney P. "The Condition of Education In The Nation." *American Education,* Vol. 7, No. 3 (April 1971), pp. 3-5.

McLendon, Jonathan C., and Haskew, Laurence D. *Views of American Schooling.* Glenview: Scott, Foresman and Company, 1964.

Miller, Richard I. *Education In A Changing Society.* Washington, D. C.: National Education Assn. of the United States, 1963.

Mitzel, Harold E. "The Impending Instruction Revolution." *The Phi Delta Kappan* 51 (April 1970): 434-439.

Morphet, Edgar L.; Johns, Roe L.; and Reller, Theodore L. *Educational Organization and Administration.* Englewoods Cliffs, N. J.: Prentice-Hall, Inc., 1967.

Morphet, Edgar L., and Ryan, Charles O., eds. *Implications For Education of Prospective Changes In Society.* Denver: Bradford-Robinson Pub. Co., 1967.

Ovard, Glen F. *Administration of the Changing Secondary School.* New York: The Macmillan Company, 1966.

Padover, Saul K. *The Genius of America.* New York: McGraw-Hill Book Co., 1960.

Paulsen, Robert F. *American Education Challenges and Images.* University of Arizona Press, 1967.

Ragan, William B. "Glimpses of the Future." *School and Society,* March 1970, p. 183.

Rankin, Stuart C. "Educational Goals in a Pluralistic Society." *Educational Leadership,* March 1971, p. 576.

Rogers, Everett M. *Diffusion of Innovations.* New York: The Free Press, Glencoe, Ill., 1962.

Saylor, J. Galen, and Alexander, William M. *Curriculum Planning for Modern Schools.* New York: Holt, Rinehart and Winston, Inc., 1966.

Schwebel, Milton. *Who Can Be Educated?* New York: Grove Press, Inc., 1968.

Shane, Harold G. "What Will the Schools Become." *Phi Delta Kappan,* June 1971, p. 597.

Sistrunk, Walter E., and Maxson, Robert C. *A Practical Approach to Secondary Social Studies.* Dubuque, Iowa: Wm. C. Brown Company Publishers, 1972.

Trump, J. Lloyd, and Baynhan, Dorsey. *The School of Tomorrow: The Teacher and The Taught.* New York: Dell Pub. Co., 1963.

Tyler, Ralph W. "Purposes for our Schools." *NASSP Bulletin,* Vol. 52, No. 332 (Dec. 1968), p. 11.

Tyler, Ralph W. "Education Must Relate to a Way of Life." *Educational Digest,* Oct. 1970, pp. 8-11.

Wahlquist, John T.; Arnold, William E.; Campbell, Ronald R.; Reller, Theodore L.; and Sands, Lester B. *The Administration of Public Schools.* New York: The Ronald Press Company, 1952.

Wey, Herbert W. *Handbook for Principals.* New York: McGraw-Hill Book Company, 1966.

Wiles, Kimball, and Patterson, Franklin. *The High School We Need.* Washington, D.C.: Association for Supervision and Curriculum Development, 1965.

HOWARD J. DEMEKE

Theory In Educational Administration: A Systems Approach

Presenting, in a single chapter, the use of theory in educational administration based on a systems (or general systems) approach proved to be no mean task. In fact it has not only proved to be difficult, but the assignment was perceived to be further from completion when the chapter manuscript was submitted to satisfy the editor's deadline than when the clear outlines of the paper first sprang from the confident pen of the writer shortly after he accepted the assignment.

Among the problems that arose during the course of trying to meet the challenge was the obvious necessity for defining a number of terms, some of them almost agonizingly complex or otherwise difficult to pin down. Worst of all were the ones, like theory itself, which submitted to no simple definition and, indeed, had to be continuously viewed through the several definitions ascribed to the term by others. Nevertheless, using a theoretically-based system consisting of perceiving, deciding and acting, the writing proceeded. In 1958 Andrew Halpin decried the repetition of (theoretical) mistakes in the early developmental days of administrative theory, particularly theory in educational administration. He cited the oft-repeated "error" of equating taxonomies with theory, as he also implied that the ranks of theoreticians had become increasingly sophisticated, thus inferring that well-meaning aspirants to membership must henceforth come forward at their own peril.[1]

---

1. Andrew W. Halpin, *Administrative Theory in Education* (Chicago: The Midwest Administration Center, University of Chicago), p. 9.

To begin with, a theory is a belief, policy, or statement of procedure proposed as the basis for action. When John Dewey promoted "learning by doing" he was undoubtedly striving to establish a principle or plan of action based on the theory that people learn better by being exposed to a variety of related experiences, including actual involvement in living (or life) experiences, than by simply reading about, or having such experiences explained to them. The origins of the administrative internship, now common in preparation programs for school administrators, had its roots solidly planted in the same theoretical soil.

It would probably be reasonable to state that the whole field of administrative theory is laced with some confusion and frustration due to not one, but several factors. In the first place, as previously indicated, the term "theory" itself has defied being subjected to any singular or simple definition. However stated, the concept of theory embodies an elusive mystique.

Even among those who can accept the definition of theory as a number of assumptions and principles which lend themselves to use in predicting or accounting for events with an accuracy better than chance,[2] they must still face the reality that terms like "assumption" and "principle" demand further clarification before universal understanding, much less agreement, is possible. After all, abstractions—frequently elusive, vague, or highly complex—are the stuff of which theoretical principles are made. These abstractions must first be reduced to concepts before hypotheses can be developed in order for selected aspects of theory to be tested.

Other terms that can be found in the literature dealing with administrative theory are words like organization, leadership, administration, educational administration and administrator. The fact that philosophers and researchers have dotted the literature with writings devoted to theoretical aspects of such terms has greatly expanded our perception of the nature and scope of these concepts. Such knowledge has also led to a greater awareness that much more research, teaching and learning will be required before theoreticians can be described as having achieved a state of open two-way communication with practicing school administrators.

It has been repeatedly suggested in the literature that the field of educational administration could greatly increase the supply of usable knowledge for intelligent decision making by becoming more theoretically oriented. If theory does indeed work in the sense that it helps to explain events in a wide number of situations, and also demonstrates a usefulness

---

2. Andrew W. Halpin, *The Leadership Behavior of School Superintendents* (Columbus, Ohio: College of Education, Ohio State University, 1956), p. 64.

as a predictor of events, then theory *is* practical ... a tool that an administrator can ill afford to be without.[3]

However, before a tool can be put to use, it is necessary for the user to first get a firm grip on it. When an administrator reads statements like: "A theory is a deductively connected set of laws. Certain of the laws are the axioms or postulates of the theory (sometimes called assumptions),"[4] he is apt to start reaching for another kind of tool. Understandably he reasons that he is a practical man, and thus he must look for tools that present greater promise of utility in the solution of problems currently confronting him.

The administrator has neither the time nor inclination to conduct in-depth research in the field of theory, but he is always on the lookout for findings and suggestions that might readily lend themselves to direct, simple, or creative application in his professional day-to-day experience.

One major purpose of this chapter, then, is to attempt to enrich the understanding of the busy school administrator in some useful ways; in other words, to help the administrator to enhance his understanding of administrative theory, and also list some suggestions for making "practical" applications of such knowledge.

ADMINISTRATION DEFINED

Administration, including educational administration, is perceived in this chapter as a process of modifying, and thus directing, the movement of an organization or group. Furthermore, administration is perceived as a process omnipresent in any organized social group, and thus, it conditions and reflects the net results of human behavior within the enterprise. This behavior includes administrative behavior which may be generalized, but probably cannot be reduced to specific cause and effect relationships.

Most of us have observed that certain specific behavioral acts "work" for some people, but will predictably fail to achieve the desired results when attempted by others. This is due to the human condition and the coincidence that administrators are also human beings. Much of the literature dealing with the story of human affairs supports the assumption that people are individually different in many ways, and will continue to manifest symptoms of this inherent human quality so long as men possess the slightest will to have a hand in the direction of their own lives.

3. *Educational Administration: Selected Readings*, ed. W. J. Hack and others (Boston: Allyn and Bacon, Inc., 1965), p. 55.
4. Daniel E. Griffiths, "Administrative Performance and System Theory," *Administration Theory* (Austin: The University of Texas, 1961), p. 7.

SYSTEMS

All organizations, not unlike people, possess similar mechanisms which govern their behavior and direction. These mechanisms are sometimes referred to as systems, which Hall and Fagan describe as "a set of objects together with relationships between the objects, and between their attributes."[5] Or, as Griffiths has more simply defined a system: ". . . a complex of elements in mutual interaction."[6]

Systems may be open or closed. In this chapter, we will be giving primary consideration to open systems—those that exchange energy and information with their environment. We shall be concerned with "feedback" processes as a dimension of the system output which, when fed back as input, becomes available for evaluation of—and modification of—succeeding outputs. Thus an open system has the property of being able to adjust future organizational direction and conduct according to past performance.[7]

LEADERSHIP DEFINED

Leadership, including educational leadership, has been variously defined. Carter would have us define this in terms of "any behaviors which experts in this area wish to consider leadership behavior."[8] On the other hand, Gibbs defined the concept somewhat differently when he indicated that leadership, when viewed in relation to an individual's leadership, is not an attribute of personality, but a quality of his role within a particular and specified social system. He added: "viewed in relation to the group, leadership is a quality of its structure."[9]

Until somewhat recently the terms leadership and administration were used interchangeably more often than not. In recent years the credibility of the distinction set forth by Lipham has gained wide acceptance, permitting us to discuss these important terms using a common frame of reference. He suggested that leadership implies the initiation of new struc-

---

5. A. D. Hall and R. E. Fagan, "General Systems," *Yearbook of the Society for the Advancement of General Systems Theory,* ed. L. V. Bertolauffy and A. Rapport (Ann Arbor: Braun-Brumfield, 1956), p. 18.

6. Daniel E. Griffiths, *Administrative Theory* (New York: Appleton-Century-Crofts, Inc., 1959), p. 9.

7. James Lipman, "Leadership and Administration," *Yearbook of the National Society of the Study of Education, 1964* (University of Chicago Press, 1964), chapter 6, p. 117.

8. Jacob W. Getzels, James M. Lipham, and Roald Campbell, *Educational Administration as a Social Process* (New York: Harper and Row, 1968), p. 24.

9. Griffiths, *Administrative Theory,* p. 68.

ture and practice to the organization or enterprise. He further suggested that administration has to do with maintaining the existing or on-going structure and/or practice.[10] Leadership, then, implies the inducement of change, while administration facilitates functional stability of the status quo.

It is important that these distinctions be clearly understood, particularly in a period when the expectations of school administrators are rapidly shifting and mere survival can no longer be easily equated with success. It may no longer be society's wish that the schools simply do better what they have long been doing anyway. Stability and smoothness of operation in the organization still give cause for admiration but, now that the schools are being forced to participate in the search for solutions to some of society's most critical problems and issues, these symptoms of administrative practice may no longer be enough. Progress is being demanded—and this can only come through change.

The problems, expectations of the day and pressures from many sources force the conclusion that change—perhaps drastic change—is being demanded of our institutions. And change requires leadership with a capability far beyond simple, though agile, reactive response to pressures. It implies the capability of initiating organizational movement towards continuous renewal and development in response to the changing needs of the society, or "publics" the organization allegedly serves. Undoubtedly one of the most troubled figures of our day is the school principal who verbalizes his own recognition of the need for, and a commitment to change, but who recoils from assuming the responsibility for educational leadership which is essential if change is to be realized in an orderly manner. He finds more personal security in demeaning but less demanding activities.

Leadership has been variously associated with concepts such as "initiating," "process," "structure," "accomplishment" and the leader's attitude and actions with respect to the relationship between the goals of an organization and those of the individuals who make up the organization. While it is understandable that one may find it inconvenient, even undesirable, to accept the exclusivity of administration and leadership, a decision to apply separate definitions to these terms was made in order to better serve the major purposes of this chapter. That such a decision (or distinction) has to be justified may serve the purpose of clearly pointing out how very limited is the extent of our real knowledge of leadership in organizations.[11]

10. Lipham, "Leadership and Administration," pp. 119-141.
11. *Ibid.*

ORGANIZATION DEFINED

What is an organization? It might be defined as a social system which includes a number of persons who perform differentiated tasks that are somewhat related due to the purposes, governance, and administration of the enterprise. Organization might also be called a process or a pattern in which a complex of related features is implied.[12] These would include a purpose (or goals), activities designed to achieve or carry out that purpose, the integration of jobs which are coordinated through a system of command, and processes . . . such as deciding, controlling, rewarding, communicating (which are crucial). In addition, there would be the emergent and prescribed motivations, attitudes, values and interactions of the organization.[13]

Although the practice of oversimplification deserves little praise, it is sometimes helpful to classify ideas, theories, or schools of thought in order to better deal with them intellectually, even though they may be overlapping and/or interdependent.[14] Thus, we can identify types of organizations, such as:

1. The "autocratic" which sought (or seeks) universal principles of management, ignorant of or ignoring the potential importance of members in favor of structure, process and management's omniscience and omnipotence in organizational understanding and direction.
2. The "human relations" school which stresses the human and group factors in organizations and fathered participatory supervision and management.
3. The "systems" or "social-system" group whose theorists concern themselves with the total system giving emphasis to the relationship of the parts to the whole, the interdependence among the range of variables, and the existence of sub-systems.
4. The "scientific"-mathematical, computer-oriented school of decision-makers undoubtedly has created a new type of organization, even though the scope of organizational decisions has been undoubtedly limited by the means utilized to reach them.

It would, of course, be possible to identify other types of organizations, not only according to their implied theories of management or decision making, but in accordance with wholly different criteria as well.

---

12. Edmund P. Learned and Audrey T. Sproat, *Organization Theory and Policy* (Homewood, Illinois: Richard D. Irvin, Inc., 1966), p. 2.
13. *Ibid.*, p. 20.
14. *Ibid.*, p. 36.

THE HUMAN DIMENSION OF ORGANIZATION

Roethlisberger confessed a deep and abiding interest in the process of communication within the organizational-administrative setting. He wondered what takes place when people interact as they addressed themselves to a common task. What, he wondered, did *they* perceive was taking place?[15]

He inferred the need for the organizational executive to think about the interpersonal proceedings in which he is engaged, and he wondered about identifying skills that the administrator could bring into practice which would thereby enable him to be "more effective as an administrator of people."[16] He also indicated that as far as he knew, there existed no single body of concepts which as yet described *systematically* and completely all the important processes that our separate theories have indicated are taking place, and how they relate to each other. He further observed that "interpersonal proceedings, unlike the atom, have not been as yet 'cracked' by social science."[17]

Efforts to "crack" or make a breakthrough, to better understandings of one's own feelings and perceptions, along with those of others, have given rise to many so-called systematic approaches. One of the best known is the process of "sensitivity training" which according to Weschler, "attempts . . . to close the gap between knowing and doing by exposing the participants to both the intellectual and emotional understanding needed for performance."[18]

There is everywhere evident the need for systems of management or administrative practice that will enable organizations to act to create their own future, rather than to continue to attempt to simply react to the pressures and uncertainties of the moment.[19] In order for an organization to free itself—to consciously develop, adopt and implement policies and procedures to effect orderly change—it becomes necessary to seek a basic change in the style of organizational leadership.

A primary philosophical assumption of this chapter is that educational administrative leadership consists of much more than waiting for

15. F. J. Roethlisberger and W. J. Dickson, *Management and the Worker* (Harvard University Press, 1946), p. 423.
16. *Ibid.*
17. *Ibid.*, p. 438.
18. I. R. Weschler, M. A. Klemes, and C. Shepard, "A New Focus on Executive Training," in *The Study of Leadership*, ed. C. G. Browne, and Thomas S. Cohn (Danville, Illinois: The Interstate Printers and Publishers, Inc., 1958), p. 441.
19. Louis A. Allen, "A Unified Theory of Administration," *Administrative Theory* (Austin: The University of Texas), p. 34.

problems to appear before acting; much more than the reactive response and token approach to change which pervades education today. The type of educational administrative leadership we are referring to implies a commitment to, and a capability for, setting in motion a state of on-going organizational change, in order that the institution shall become adaptive to the changing needs of the learner and the demands of the society in which he lives.

Andrew Halpin found in studying a number of superintendents that the school board, staff, and teachers felt the superintendent's most serious deficiency was ineffective leadership. The study concluded that the superintendent erred most because he failed to provide the school with an organization that met the requirements of "initiating structure."[20]

Principles of administration were traditionally believed to have utility, at least to the point where they clashed with the empirical realities of behavior. Today one trend is to attempt to view administrative functions rationally. Administrative behavior, as a field of study in organizational interrelationships, is believed by some theoreticians to be subject to description and definitive analysis. There exists widespread but largely unspoken disdain for prescribed techniques in administration, based wholly on value judgments unsupported by "scientific" evidence.[21]

More than fifteen years ago Walton noted that the then current range of theories appeared to be symptomatic of an underlying dissatisfaction with traditional principles and techniques in administration.[22] What was shortly thereafter perceived to be needed is a theory of administrative behavior that will provide a set of guides to action and fact collection, that will also describe the nature of administration and provide a framework within which researchers can find new knowledge.[23]

Theories of administration in a number of fields, including business, industry, public administration and education, are similar in that they are concerned with the concept of getting the job done with due regard for the morale of the personnel.[24]

20. Halpin, *Leadership Behavior of Superintendents*, p. 118.
21. Leonard Feroleto, "An Analysis of Educational Leadership in a Democratic Setting" (Doctoral Dissertation, Arizona State University, 1969), p. 14.
22. John Walton, "The Theoretical Study of Educational Administration," *Harvard Review*, 1955, p. 69.
23. Daniel E. Griffiths, "Administration as Decision Making," in *Administrative Theory in Education* (Chicago: The Mid-West Administrative Center, University of Chicago, 1958), p. 119.
24. Richard W. Saxe, "The Principal and Theory," in *Perspectives on the Changing Role of the Principal* (Springfield, Illinois: Charles C. Thomas, Publisher, 1968), p. 5.

Probably the basic act performed by the administrator is decision making. Primary decisions are made to influence the formulation of policies which determine an organization's function and performance. Subsequent decisions, which are sometimes referred to as operational decisions, are actions designed to facilitate the achievement of goals governed by policies. These task-oriented decisions are made and implemented by administrators.

If all kinds of theories of administration are, indeed, similar to educational administration theories, in that they revolve around the concept of getting the job done and keeping the "troops" happy, the solution to many administrative problems in education might well be discovered through the study of administrative theory and practice unrelated to education.

By way of illustration, let us pursue the subject of administrative theory by utilizing the concepts of effectiveness (getting the job done) and efficiency (preserving personnel morale) as formulated by Chester I. Barnard. The former refers to the achievement of the cooperative purpose, which is essentially nonpersonal in character. The latter refers primarily to the satisfaction of individual desires, which is highly personal in character.[25]

The aforementioned theory suggests, for example, that one way to successful administration is through the development of a thorough understanding of the relationships of these two concepts, together with the realization that they must "balance" according to the administrative setting. The administrator will thereby (theoretically) possess an adequate understanding, or frame of reference, with which to guide his actions and thus confidently and successfully fulfill his role.[26]

Another, but related, theory holds that, to be successful a school administrator—any school administrator—must demonstrate a working knowledge of the premise that each person is primarily interested in his own goals; not the objectives of others. Individuals will work willingly and productively for the goals of the enterprise only when they see a satisfactory return for themselves from the group effort.[27] In other words: "People join together in groups to work for common goals because the pooled efforts of many produce greater returns for each, than the uncoordinated efforts of individuals."[28] Saxe has stated it this way: "It seems

---

25. Chester I. Barnard, *The Functions of the Executive* (Cambridge: Harvard University Press, 1964), pp. 41-48.
26. Saxe, "Principal and Theory," pp. 5-7.
27. Allen, "Unified Theory of Administration," p. 21.
28. *Ibid.*, p. 33.

that a one-sided dedication to the institutional goal of providing a good education for clients is not sufficient motivation for teachers."[29]

All administrators operate on the basis of theory whether they know it or not. Decisions are commitments to act. Decisions must be made and are being made every day. Even a decision to avoid making a decision is a decision to act in a given way. In every case a decision is made by choosing from among perceived alternatives. Awareness of the probable consequences, or the ability to predict the probable outcomes of decisions, are usually seen as a reflection of the administrator's past experience. Certainly the administrator reaches into his own past in order to exercise judgment in current decision making. But what does the administrator find when he searches his store of knowledge and experience? Chances are that his mental "tour" will isolate fragments of randomly accumulated knowledge gleaned from many largely isolated, unrelated or limited experiences. From this mass of often conflicting, frequently vaguely remembered data, he must select among the perceived options something that is in harmony with his own values. This option—a way of acting—hopefully will also work in the current situation. It is suggested that such an approach, although still widely employed among educational administrators, could be improved, perhaps vastly improved, were the practitioner to develop a better understanding of the meaning and usefulness of theory in his professional life.

A list of the major purposes to be served through embracing theoretical approaches to administration might include the following:

1. To give rational meaning and order to otherwise only slightly related or randomized knowledge.
2. To assist one to better conceptualize the relationships that exist between organizational components, people and/or events.
3. To apply ordered intelligence to the end that the results of administrative/organizational decisions become increasingly predictable.

In other words, the aim is to encourage and assist the administrator to gradually develop a more practical—workable—theory of administration in place of the relatively unsophisticated approaches he now uses as bases for his decisions and behavior.

The goal, therefore, is not to discover *the* theory of administration, but to work gradually, inexorably and intelligently toward a higher level of understanding of the meaning of theory and a deeper appreciation of its increasing usefulness to the administrator who will work diligently to

---

29. Learned and Sproat, *Organization Theory and Policy*, p. 7.

delineate and then continue to develop and refine a "personal" theory of educational administration.

One way to get at this task is through the use of a systems approach. For the sake of illustration, let us imagine a school principal who is desirous of having his faculty develop a working knowledge of a new process and procedure in the teaching of reading. He must somehow develop a theoretical foundation upon which he can base a series of decisions calculated to transfer his ideas from the relative remoteness of theory to the reality of action.

The principal might hypothesize that acquisition of the skills and knowledge he proposes to have introduced will, when acquired, actually modify each teacher's behavior and thereby improve each individual's professional practice in the teaching of reading. He must also develop from certain theoretical assumptions he holds regarding the administrative process a related set of hypotheses to be used as additional bases for action. Saunders, et al. refer to this as the "if-then" stage, as opposed to the theoretical stage which is used to explain the "why" of our actions and behavior.[30]

The number of possible alternative approaches to the solution to this administrator's problem cannot be exactly determined, but would obviously be many. The principal, however, must decide on a course of action, hypothesizing that certain desired results will occur if he makes certain appropriate decisions and then acts appropriately to implement them. Among possible differences between the use of a systems approach to decision making and a number of other approaches, it is the insistence on first identifying a clearly defined goal (or goals) before beginning to plan and schedule a course of action, plus the allocation of resources required to support the mission. Provision for regular feedback and evaluation are also basic elements. Systems therefore operate within a conceptual framework or model.

In the case cited, the principal put himself through a multi-step process designed to include, but not be limited to the following steps:

- He assumed (theorized) that he had acquired new insight into the existence and availability of a fund of learning activities related to the teaching of reading.
- He then further hypothesized that, if this knowledge was acquired by members of the teaching staff, it would result in improved

---

30. Robert L. Saunders, Ray C. Phillips, and Harold J. Johnson, *A Theory of Educational Leadership* (Columbus, Ohio: Charles E. Merrill Books, Inc., 1966), p. 18.

educational opportunities for students. (This is a major goal.) He must then further theorize regarding the best possible set of conditions conducive to achieving the goal(s) intended.

- He then selects a basic approach, including a general framework of conditions which, hypothetically, will enhance the chances of success of the enterprise. In this case the principal must base his decisions on his perceived knowledge of how people learn. These decisions will be in harmony with his own value system regarding the relative merits of various supervisory-inservice (psychological) approaches. Hopefully, he would be aware that his objectivity might be somewhat blurred by his "closeness" to the situation and/or various pressures of the moment, and that his resources, including human resources, have limitations of which he is also aware.

- Once he has decided on a basic approach ("learning by doing," for instance) he is then in a position to develop a system, based on total staff involvement, designed to achieve the pre-determined goal(s).

- He proceeds to involve staff members, assuming that teachers consciously desire to improve their professional practice and will therefore be motivated to lend their support to the project once they perceive that it contains something they seek for themselves.[31] The resulting program, ostensibly designed by the school staff (input), is implemented, but not before building in a mechanism for generating and receiving appropriate feedback to be used to facilitate regularized system modification. The latter is uniquely important in a systems approach because it permits the enterprise to promptly respond to internally or externally generated stimuli; thus raising appreciably the chances for the mission's success.

- The "moment of truth" comes, of course, when the results are in from testing the hypotheses. It is at this stage that success or failure, or degrees thereof, are determined in the cold light of reality; where appraisal can be deeply rewarding, agonizingly disappointing, or somewhere on the continuum between these two extremes.

The school principal has, in a sense, moved a set of ideas (philosophical assumptions, scientific principles, theoretical generalizations) from the theoretical stage to the reality stage. Theory has thus resulted in hypoth-

---

31. Allen, "Unified Theory of Administration," p. 21.

eses which have been translated to action. By developing hypotheses out of the stuff of his own theoretical base, and then subjecting them to investigation, including actual testing, he has systematized the utilization of theory in the governance of his own and the organization's actions. Although his every action might have been rooted in a consistent theoretical base, the administrator has not been constrained to tie his decisions to any particular or unique theory of administration. On the contrary, lacking access to a viable theory of educational administration, he has behaved eclectically, through selecting from his perceived knowledge of many theories. He has also elected to give greater weight to some theoretical assumptions and generalizations than others. As Griffiths has stated: "theory is useful to a researcher because it gives him a benchmark from which to start, not a mooring against which to rest."[32] Could we assume that, lacking a viable theory, the same position is a most defensible one for the administrator?

Griffiths' statement could probably serve to describe the way most school administrators attitudinally approach the use of theory in their daily work. Then there is also the temptation to speculate whether his statement represents the best of theories of administrative behavior. To go one step further, it is seriously suggested that the real rewards to be found in the study of educational administrative theory might well be viewed as by-products of an endless search for certain useful dimensions of existing theories, and of those to come.

Theories gain respect when hypotheses growing out of them stand the tests of investigation and evaluation.[33] The quality of such investigation is enhanced, in most cases, where certain variables are held constant. When theories are subjected to testing in practice, the quality of the generalizations that can be derived from observing results are often suspect because of the researcher's inability to control one important variable—the administrator himself. School administrators are people; people who are alike in some ways and wonderfully and dependably different in many more.

Could it be that an abiding and irrefutable axiom of educational administration is that the administrator must, because of being the unique person he is, set his own style? Theories offer choices, but the choosing will be done by, and will differ with each administrator. So the search for theory should continue; not to find *the* theory, but to add to the choices of tested options open to administrators. We are reminded that administrators will each inevitably be operating at different levels of perception

32. Griffiths, *Administrative Theory*, p. 5.
33. Halpin, *Administrative Theory in Education*, p. 59.

38 THEORY IN EDUCATIONAL ADMINISTRATION

and sophistication within organizations, no two of which are ever likely to be identical. Moreover, each organization functions within a unique social milieu; as different perhaps as the individual administrator himself who strives to give direction to the organization he serves. Students interested in pursuing this line of thought are encouraged to seek other sources, but particularly, Chapter 1 of *Human Relations in School Administration*,[34] where Griffiths discusses the "Tri-dimensional," (The Man—The Job and The Social Setting) Concept and the three skill (technical-human-conceptual) approach, giving appropriate credit to Daniel Davis, Ernest Weinrich and Robert L. Katz for their contributions.[35] Essentially, taxonomic rather than theoretical in structure, the tri-dimensional concept can help the school administrator to obtain a clearer understanding of his own professional and personal uniqueness, as well as his administrative role in the social setting in which he finds himself. Delitescent rewards await the administrator who becomes a student of theory. After all, can any administrator claim to have a continuing and unfailing gift for serendipity?

Lacking a theoretical framework from which to operate, school administrators are often prone to failure in their attempts to clearly identify and analyze problems that confront them. Not only is there the danger of failure to understand problems, there is the larger danger of not being able to see some problems at all—or even identifying the wrong ones.[36]

One of the purposes of theory is certainly to provide improved awareness regarding possible consequences which might result from alternative decisions. Decisions by the administrator require him to make choices from among options, or perceived alternatives. His own value system and his unique experiential background are brought into play as he reaches his selection of the "best" decision open to him.[37]

Theory, then, gives the administrator a tool for the analysis and discrimination of situations which may otherwise be misleading, or go unrecognized until there is no longer adequate time for appropriate positive action.[38] Part of the problem may stem from the possibility that much of the knowledge usually employed by administrators, while unquestionably important to each as an individual, has been randomly acquired through his own experience and is not suitable for use in drawing generalizations.[39]

34. Daniel E. Griffiths, *Human Relations in School Administration* (New York: Appleton-Century-Crofts, Inc., 1956), pp. 3-22.
35. *Ibid.*, p. 4.
36. Halpin, *The Leadership Behavior of School Superintendents*, p. 64.
37. *Educational Administration: Selected Reading*, ed. Hack and others, p. 55.
38. Halpin, *The Leadership Behavior of School Superintendents*, p. 64.
39. *Educational Administration: Selected Reading*, op. cit.

It has been suggested that much of what has traditionally passed for administrative theory has been derived from what practitioners have said they do, or have done. The difficulty faced by the researcher in his efforts to obtain reliable data via this route has been compared with the relatively easy time enjoyed by researchers who choose to determine how atoms and viruses behave. Coladarci and Getzels cite the advantages inherent in the fact that such things do not talk.[40]

In any event the search for an adequate theory of educational administration goes on. The pursuit has gained in popularity, particularly in the last two decades, and this can only add to our understanding of a field that encompasses management, organization, and leadership as well.[41] The need for more reliable guidelines or principles of modern administrative performance is recognized today. Lacking any foreseeable alternatives, administrative theory currently becomes the most promising potential source of this needed direction.

Although we cannot foresee the problems that will be confronting the educational administrator in the years to come, it can be predicted that values will continue to change along with the conditions surrounding them. Somehow a way must be found to adequately equip the school administrator to wisely alter the values of those variables within the scope of, and subject to, his control when other, perhaps increasing numbers of variables beyond his control change in value.[42]

Failure in this mission can predictably mean continuing frustration, certain obsolescence and perhaps failure and, ultimately, elimination of the professional administrative role in education. This possibility becomes more readily apparent if we recognize and accept the assumption that a major function of administration—and the administrator—in education is to improve, not simply maintain, the productivity of the educational profession and the establishment it serves.

There are, admittedly, too many different meanings ascribed to the term theory.[43] However, if we can think of the search for theory, or theory development, as a process involving a method of thinking, reasoning and problem solving,[44] then the possibilities for combining a systematic approach with theory become more apparent. Viewed as a process, theory

---

40. Arthur P. Coladarci and Jacob W. Getzels, *The Use of Theory In Educational Administration*, Monograph No. 5, School of Education, Stanford University, Stanford, California, 1955.
41. Saunders et al., *A Theory of Educational Leadership*, pp. v, vi.
42. *Ibid.*
43. Halpin, *Administrative Theory in Education*, p. 6.
44. Saunders et al., *Theory of Educational Leadership*, p. 6.

lends itself to convenient comparison with the process of educational administration.

Administrative theory when perceived as process encourages and promotes the postulation of models of administrative thought and action from which hypotheses can be drawn and tested. Such an approach also is conducive to the objective observation and description of events, apart from any evaluation which could and should be conducted separately.[45] Viewing administrative theory as a process permits us to concern ourselves primarily with cause and effect: what the administrator does and/or causes to happen, rather than valuing. Furthermore, any predictions resulting from the testing of related hypotheses can be based on "recognition of the fact that human behavior flows out of persons in situations."[46]

If administration is to become a scientifically oriented discipline, and the aim of science is to predict, it follows that the discipline shall include the predictability of behavior in the organizational setting. However, there are many of us who presently believe that human behavior does not always reveal itself in purely rational ways; who believe that man does not usually behave in accordance with any scientific law. Men do develop habits, but even these are not always practiced with consistency. Therefore we say that man's behavior frequently defies prediction.

Nevertheless we have some cause for hope that behavior may be predicted with increasing dependability. Man has developed a hierarchy of objectives which might be referred to as his system of values. There are many commonalities in men's choices of goals, although such goals may be pursued in diverse ways, guided by values extending from the goals.

For example, let us explore what is called role theory in educational administration. Based on a number of assumptions, including one that holds that only members of a given profession are qualified to determine the areas of competence that must be successfully assumed by the complete professional, a list of necessarily interdependent roles is circumscribed. The respective areas of competence (or roles) are described, sometimes in analytical form, as behavioral activities amenable to direct observation, or indirectly through verification of one form or another, including skillful interview.

The following role description and analysis may serve to illustrate how role theory has been used to develop a behavioral definition of the

---

45. Halpin, *op. cit.*, pp. 5-7.
46. James O. Thompson, "Modern Approaches to Theory in Administration," in *Educational Administration: Selected Readings,* ed. Hack and others (Boston: Allyn and Bacon, Inc., 1965), p. 97.

school administrator's role, or "areas of competence," in personnel administration.[47]

The distinction between "competence" and "effectiveness" should always be kept in mind with respect to role theory. Competence refers to the appropriate use of a body of knowledge and skills which members of a professional group hold privy to their group. Effectiveness refers to the long and short range effects of administrative actions. Role theorists generally subscribe to the position that the administrator who "masters" certain areas of competence (roles) will predictably behave more effectively than those who do not so learn or subscribe.

To the role theorist in medicine, for instance, a surgeon could demonstrate his professional competence yet still lose a patient. In other words, one can be competent while also being ineffective.

Note that the entire role is (1) described and (2) analysed in terms of what the administrator does in the course of "competently" assuming one of a number of roles assigned to him.

Administration of Personnel (Role of Area of Competence)*

Normally the administrator will participate, with due preparation, in the selection and assignment of faculty and staff personnel. In so doing, the ethical administrator accepts a large measure of responsibility for on-the-job success of such personnel. He will participate with them in a careful program of evaluation of their individual competence and then work diligently to assist each person to reach ever-higher levels of performance in areas of desired competence. He recommends personnel for retention or termination on the basis of more-than-adequate written objective data. He fully recognizes that a recommendation to terminate a staff member is sometimes necessary, but he also acknowledges that a part of such failure is due to his own ineptitude in personnel administration.

---

47. Howard J. Demeke, *Guidelines for Evaluation: The School Principal—Seven Areas of Competence* (Tempe, Arizona: University Book Store, Arizona State University, 1971), pp. 51-56.

*Note: "Administrator of Personnel" is one of seven "areas of competence" described and analyzed in a publication by the author. It was designed to assist administrators, particularly school principals, desirous of engaging in a process of self-evaluation leading to improvement of their professional practice. The others areas of competence are: (1) Leader and Director/Facilitator of the Educational Program, (2) Coordinator of Guidance and Special Services, (3) Link Between the School and Community, (4) Member of the Profession of Educational Administration, (5) Member of the District and School Staff, and (6) Director of Support Management.[48]

48. Demeke, *Guideline for Evaluation.*

42 THEORY IN EDUCATIONAL ADMINISTRATION

In this role, the competent administrator:

(Area Description)

- Develops skillful techniques related to selecting, assigning, assisting, and evaluating the competence of personnel. He utilizes both formal and informal procedures—but these are useful only because he has established appropriate and effective personal and professional relationships.

- Recognizes that self-evaluation carries the greatest promise for engendering productive behavioral change in the individual, thus he gives continuous support and encouragement to this approach.

- Assigns (and reassigns) staff members to positions in harmony with their prior experience, education and demonstrated competence. He believes that preventing a staff member from having to serve in a position where he lacks competence is at least as important to the program's success as the appropriate assignment of personnel. Perhaps even more so.

- Uses the school district's definition of professional competence as the criterion in making his recommendations. If for any reason the district does not have such a definition, he strongly urges and assists in its development and ultimate adoption. This would, of course, include "spelling out" the areas of competence of the administrator, as well as those for teachers and others.

Administrator of Personnel:

(Area Analysis)

5.1 Participates in selection and assignment of personnel
    5.11 Knows how to obtain and utilize all available information pertinent to personnel selection
        5.111 Skillfully compares such information with the personal and professional qualities desired and/or required for productive membership on the school staff
        5.112 Assigns personnel on basis of training and probable competence
        5.113 Attempts to build a balance of abilities, interests and talents within the school staff
5.2 Defines and clarifies responsibilities for all employees and potential employees
    5.21 Maintains accurate and up-to-date position descriptions for all teaching and non-teaching assignments

- 5.22 Assigns faculty and staff duties and responsibilities clearly and carefully
- 5.23 Provides a handbook of school policies and practices for use by staff members
- 5.24 Makes special staff requests or assignments well in advance whenever possible, and in writing as necessary
- 5.25 Checks to see that responsibilities, once assigned, are carried out

5.3 Handles confidential information ethically and professionally
- 5.31 Secures confidential papers to ensure that they are not available to unauthorized personnel
- 5.32 Discusses confidential matters only with those appropriately privileged to share them
- 5.33 Honors staff confidences

5.4 Provides increased time for instructional and administrative staff to study and improve program of education
- 5.41 Delegates all possible clerical and mangerial responsibilities (not policy) to others.

5.5 Accepts responsibility for the evaluation of staff competence
- 5.51 Uses district definition of professional competence as the basic criterion when developing his recommendations for retention or termination
- 5.52 Recommends retention or termination only after careful and systematic collection and evaluation of objective data
    - 5.521 Recommends for permanent status only those who demonstrate high professional competency and potential
- 5.53 Documents all recommendations for service extension or termination
- 5.54 Personally notifies personnel of intention to terminate service, when possible, at least one semester prior to terminal date
    - 5.541 Carefully identifies areas of competence requiring improvement
    - 5.542 Makes specific suggestions aimed at overcoming weaknesses
    - 5.543 Actively works with personnel in a resource-support role in an effort to facilitate improved professional competence

5.6 Uses systematic and effective evaluation procedures
- 5.61 Visits classroom regularly

44 THEORY IN EDUCATIONAL ADMINISTRATION

- 5.611 In pre-conference ascertains the teacher's instructional objectives for the lesson to be observed
- 5.612 Demonstrates skill in data collection and other observational techniques
- 5.613 Discusses data with teacher following observation
- 5.62 Demonstrates skill in obtaining data relevant to teaching competence through interviews
- 5.63 Shows skill in conducting evaluation conference
  - 5.631 Enumerates areas of strength and weakness
  - 5.632 Asks appropriate but searching questions
  - 5.633 Demonstrates his own listening skills
  - 5.634 Whenever possible is constructively critical while being positively reassuring
- 5.64 Fully accepts his share of responsibility for the success or failure of personnel, particularly those who were recommended by him for employment and/or assignment
  - 5.641 Devotes appropriate time and energy to activities conducive to staff improvement
- 5.65 Arranges to deeply involve staff members in inservice activities leading to more sophisticated skills and procedures in self-evaluation in all areas of teacher competence

Social scientists have striven to increase the probability of predictable behavior in man. What success they have achieved has probably been greatly due to their efforts to understand the hierarchy of goals perceived by men, and by becoming aware of behavioral factors identified with men striving to maintain their value system. Free choice is operative, but men do form habits which may be changed (else why educate?), but with varying degree of difficulty.[49]

Building theory is a creative thinking activity; not an expository, analytical, explanatory, or even a research function. Some individuals with the capability and disposition to think creatively have little inclination to pursue, and therefore, find little time for research investigation. They are probably challenged when they reason that if their ideas have merit, other types will probably volunteer to undertake the job of testing them. Not

---

49. Thomas Casey, "The Compatability of Free Choice With the Prediction of Human Behavior," *Rocky Mountain Science Journal*, October 1967, pp. 159.

every theoretician, of course, behaves in this way. Indeed, as a breed they appear to be anything but herd prone.

The task of the theoretician is to first develop a set of concepts that permit the description, in terms relevant to the theory, of administrative situations. These concepts must be operational in that their meanings should correspond to observable facts and situations.[50] But in every case it should be remembered that "theories will always be transitory and incomplete because of the modification in our surroundings which result from the application of new knowledge."[51] The same, however, might be said of a systems approach to administration, particularly when the latter is viewed as process.

A major theoretical assumption suggested in this chapter is that the essence of school administration is organizational renewal and development.

A second assumption is that the school administrator is—for good or ill—the catalyst who by his decisions and action largely determines the form and degree of change evident in the continuous reorientation of the organization and its components.

A third theoretical assumption is that achieving institutional change rather than, say, maintenance of the status quo, is the primary function of the professional educational administrator. This assumption is embraced, notwithstanding the possibility that both he and his several "publics" may reject such an assumption, at least temporarily. This is because there is often notable lack of agreement among school administrators, teachers serving on their respective faculties, and non-educators regarding the characteristics of the effective administrator.[52] A study made by Reed L. Buffington and Leland Medsker[53] that involved principals, teachers and parents from each of thirty schools prompted the following report.

The teachers viewed the principal's most important job as that of providing leadership for teachers. The parents placed major emphasis on the principal and his responsibility for developing effective relationships with parents' groups and

---

50. Herbert Simon, *Administrative Behavior* (New York: The Macmillan Company, 1956), p. 37.

51. Luvern Cunningham, "Research in External Administration—What do We Know," *Administrative Theory as a Guide to Action,* ed. Campbell and Lipham (Chicago: Midwest Administrative Center, University of Chicago, 1960), p. 149.

52. This statement represents a unilateral broadening of a generalization which the originators applied only to the school principal. See John K. Hemphill, Daniel Griffiths and Norman Frederickson, *Administrative Performance and Personality* (New York: Bureau of Publications, Teachers College, Columbia University, 1962), p. 339.

53. Reed L. Buffington and Leland Medsker, "Teachers and Parents Describe the Effective Principal's Behavior," *Administrators Notebook* 4 (September 1965): 1-4.

the community. The teachers viewed such relationships as important but ranked them third in importance among the principal's responsibilities. Both the parents and the teachers ranked the principal's work with, and service, to children as second in importance among his responsibilities, but the elements of such work and service were stated somewhat differently by the two groups. The parents made little reference to the principal's relationships with teachers, and neither group said anything about his relationships with the superintendent. And, finally, neither group placed any emphasis on the principal's responsibilities in the supervision of instruction or in curriculum development.[54]

An administrator will view his own behavior in terms of the expectations he personally holds for his position. The probability that he may be the only person who holds such expectations may or may not deny the importance to the administrator of having what he does and why he does it accurately perceived and accepted by those around him. The probability of rejection of the third assumption rests on the observation that much of the frustration and conflict to be found in the schools is due in large measure to variances in role expectations which individuals (including administrators) hold for themselves and for persons who occupy either different or like positions.[55]

If a theory is comprised of a set of assumptions, then a theory must also somehow suggest the methodology to be employed in testing any hypotheses growing out of the theory. It should, however, be remembered that theory demands creative thought, primarily, to give it form; the testing of hypotheses calls for other, though precise, selected, research skills.

Halpin was encouraging, where he noted the variables in the evolutionary development of theories, and further urged that: "One must respect these differences and must recognize that theories, like the human beings who create them, follow different courses of development, and grow at different rates."[56]

In this chapter a systems model has been proposed which was intended to further explain, while finding support in, the theoretical assumptions previously stated. The model was designed to provide an administrative guide to action for the collection of objective and subjective data, and for use in describing the essential nature of educational administration within the scope of the organizational renewal-development assumption. Thus an effort has been made to meet criteria suggested elsewhere as minimal for models associated with theories of administrative behavior.[57]

54. William W. Savage, *Interpersonal and Group Relations in Educational Administration* (Glenview, Illinois: Scott, Foresman and Company, 1968), p. 136.
55. *Ibid.*, pp. 120-153.
56. Andrew Halpin, *Theory and Research in Administration* (New York: The Macmillan Company, 1966), pp. 6-7.
57. Griffiths, *Administrative Theory in Education*, p. 119.

Insofar as the provision of an administrative guide to action is concerned, we should consider that, while one aim of theory is to somehow provide knowledge sufficient to permit the prediction of consequences growing out of administrative situations,[58] "prediction is always hazardous in social situations."[59]

A SUGGESTED MODEL FOR A SYSTEMS APPROACH TO ADMINISTRATION WHEN THE LATTER IS PERCEIVED AS ORGANIZATIONAL RENEWAL AND DEVELOPMENT

One frequently reads about and hears conversation referring to systems, systems analysis, systems theory, systems approach, and the like. But what is meant by the term system? In general a system might be described as a group of objects or elements that exhibit independent, yet interdependent, relationships as they operate as a unit (or sub-system) to produce or achieve a specific mission, or produce a total effect.[60]

Although a systems approach to administrative problem-identification and solution may not work in all situations,[61] the approach is certainly one of the most provocative and promising to come along in years. A concept of educational administration that incorporates the regularization of needs-assessment, constant examination of facts, beliefs, and values, plus the development, testing and evaluation of goals, objectives and results, cannot be lightly dismissed.[62]

Educational administrative theory can be used instrumentally to help the administrator to see his way more clearly through the "fog" of contrasting values, pressures and constraints. Theory can further enable him to "take his bearings" in order to establish a base for launching out in predetermined directions to meet the challenge of now-identified problems. By subscribing to procedures inherent in systems analysis, he employs the process of evaluation as perhaps the one indispensible factor that must be present before any intelligent decision—and therefore any productive progress—becomes possible. In other words, the reader is asked to seriously consider whether administrative decisions resulting from a reliance

---

58. Herbert Simon, *Administrative Behavior*, p. 149.
59. Keith Goldhammer, *Issues and Strategies in the Public Acceptance of Change* (Eugene: University of Oregon, Center for Advanced Study of Educational Administration, 18 November 1965), p. 16.
60. For two apparently closely related definitions see: Floyd H. Allport, *Theories of Perception and Concept of Structure* (New York: John Wiley and Sons, 1955), p. 649; and A. D. Hall and R. E. Fagan, "General Systems," *Yearbook of the Society for the Advancement of General Systems Theory*, p. 18.
61. *Educational Administration: Selected Readings*, ed. Hack and others, p. 377.
62. *Ibid.*

on systems theory can be appropriate for utilization in the governance of personal and organizational direction.[63] Systems theory can also be used to test the soundness and consistency of governing values and principles; also as a guide to evaluating the behavior, methods and procedures employed to reach the stated objectives—and reasons for their achievement or non-achievement, and to what degree.

Elsewhere in this chapter it was suggested that the modern educational administrator must develop a new philosophy, plus techniques which will enable him to combine in his own behavior distinctive aspects of leadership and administration as these have been re-defined. Ways must be found to institutionalize change, while enhancing the productivity of the schools. Furthermore, productive change needs to be implemented on a reasonable scale without delay, without the introduction of undo periodic conflict, or traumatic organizational instability, and within the existing framework of public education.

In the following pages are presented the broad outlines of a process designed to assist educational leaders to systematically initiate and maintain change in the organizations they serve. The proposal was perceived as a multi-phased, laboratory-oriented program and assumes the deep personal involvement of all employees sooner or later. It was designed to initially attract those employees who would voluntarily admit to a concern for finding ways of improving their personal competence and teamwork for the broad purpose of developing vehicles (systems) for effecting needed change in the institution or organization.

Major objectives of the proposed process might include, but not be limited to, the following:

1. To develop improved understanding of, and respect for, the team concept as this relates positively to personal and staff morale, and institutional/organizational productivity.

2. To involve participants in a process of institutional and personal self-evaluation in order to identify major problems contributing to the difficulty of providing adequate and relevant educational opportunities for all students.

3. To provide in-service training, education and experiences that will enable participants to demonstrate improved, hence more productive, interpersonal and group relations in their day-to-day associations with their fellow workers.

63. *Ibid.*, p. 368.

4. To explore techniques and develop understandings calculated to generate individual self-respect, plus mutual recognition, pleasure and affection among participants to the end that greater pride and mutual respect shall develop and prevail.[64]

5. To develop procedures for regularized feedback and periodic reassessment designed to sustain purposes through timely and appropriate program adjustment. This sub-system conceivably could include follow-up educational services to provide for further development of interpersonal and group relations within the total school community.

6. To promote the desirability, invite the participation, and test and evaluate the feasibility of actively involving in the organizational renewal process representatives from all segments of the school community, particularly parents and students, whose understanding and support have become imperatives for the education of American children and youth in our time.

7. To develop, test, and evaluate operational designs for the resolution of identified problems, and for dealing with issues. Special concern would be given to new or innovative approaches, practices, program content, and logistics. The inclusion of qualitative research support would be imperative here.

A number of theoretical assumptions are identified and used as foundational supports for the proposed model.[65] A few of the more important ones are in the following list:

1. Organizational, institutional or individual progress is always possible, but only through change.
2. Institutional/organizational change depends on changes in the attitudes, values and work habits of persons serving the organization directly or indirectly.

64. Griffiths refers to five components of job satisfaction previously identified by Talcott Parsons. Objective #4 above mentions four of these: Recognition, Self-respect, Pleasure and Affection. Not included: Satisfaction of Wants. in *Human Relations in School Administration*, pp. 18-36.
65. Robert L. Saunders, et al. have developed a theory of educational leadership with the emphasis on instructional improvement. Their theory rests on a number of assumptions which coincidently, but not surprisingly, appear to be closely related to a number of those listed above. The major concepts comprising their theory derive from the nature of man, the nature of decision making in a democratic society, the nature of leadership, and a method of solving problems. A comprehensive discourse is devoted to the presentation of evidence in support of these concepts (chapters 6-10, inclusive; pp. 42-122).

3. Changes in attitudes, values and work habits can be initiated externally to the individual or group.
4. Change will occur when the individual perceives that he wishes to change, knows how to change, and does so voluntarily.
5. The sum total of individual change will be reflected in resultant institutional/organizational change; this change will approximate the desired change to the degree that the underlying theory has validity and the program utilized is a reliable vehicle for sustaining it.
6. Leadership is characterized by the ability to generate a positive response due, in part, to the leader's ability to help the group's members move to a position of shared consensus.
7. Individuals and groups will accept responsibilities best when they are involved in the formulation of policies out of which these responsibilities grow.

SCOPE OF THE PROPOSED SYSTEM

The process, as perceived, would initially consist of five phases and would include provision for regular feedback, evaluation and appropriate modification intended to upgrade the system and thus its productive impact. It is proposed that the five (initial) phases include: (see Fig. 1)

Phase I — The Planning Phase — Including the development of tentative purposes and objectives, in behavioral terms where appropriate. Also included would be the development of program scope and sequence, plus the selection—at least tentatively—of techniques for evaluating the vehicle in terms of the degree to which the stated purposes are being achieved. Accordingly, tentative plans would also be developed for initiating appropriate modifications in either the program system or the organization itself, to ensure that each would be responsible to the other. All administrative-supervisory personnel who stand to be held responsible for implementing decisions growing out of the system would be represented in deliberations at some point in Phase I. Titular or defacto leaders of subgroups, formal and informal, might be initially involved at Phase I. However, the dynamics of the situation could conceivably warrant a decision to involve them at Phase II.

Whether "outside" consultants would be involved in the total process would be a decision that should be made early in Phase I. In the face of pressure to develop a methodology of educational administration which stresses democratic approaches to decision making and implementation[66] it would probably be wise to consider the values inherent in situations

66. *Ibid.*

A Model for a Systems Approach to Educational Administration in the 1970's

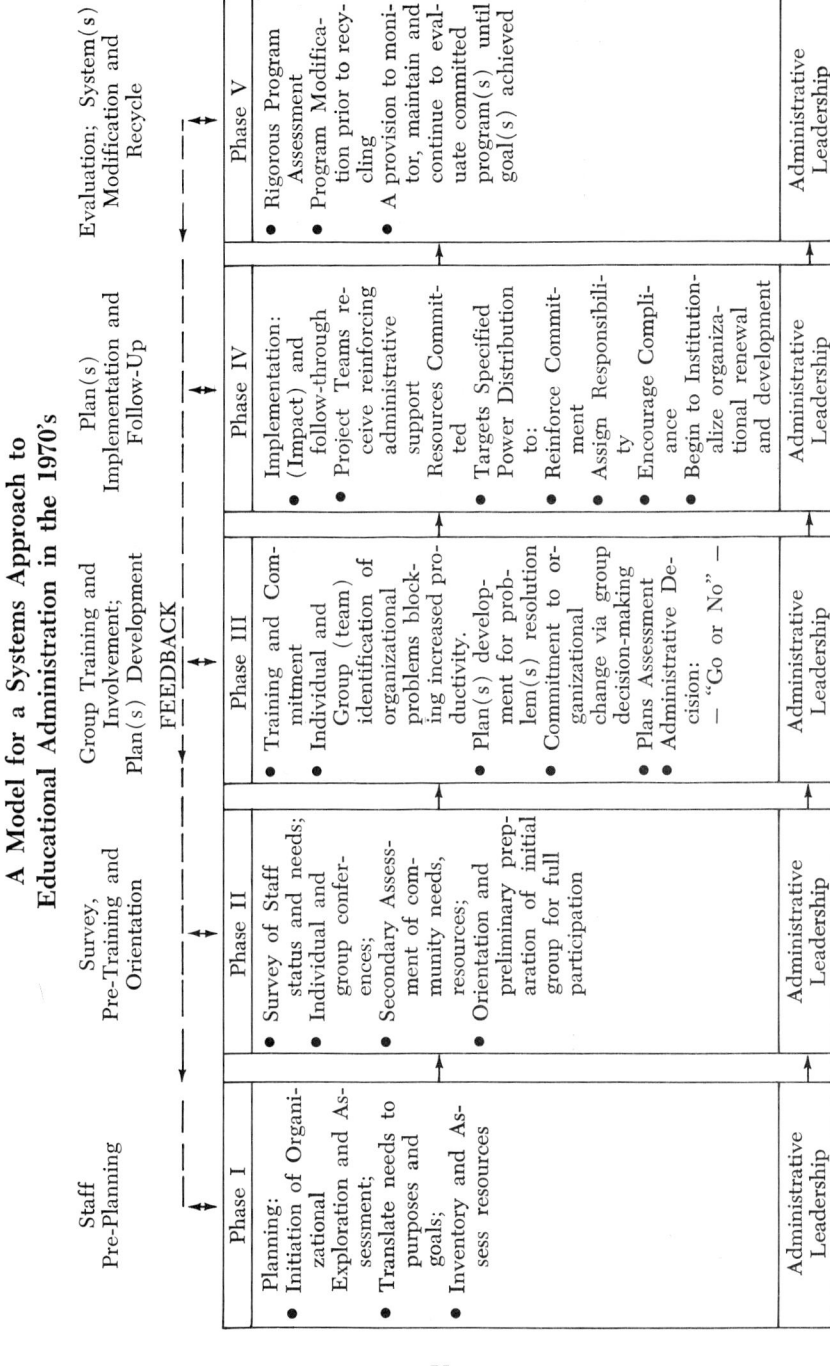

Staff Pre-Planning	Survey, Pre-Training and Orientation	Group Training and Involvement; Plan(s) Development	Plan(s) Implementation and Follow-Up	Evaluation; System(s) Modification and Recycle
		FEEDBACK		
Phase I	**Phase II**	**Phase III**	**Phase IV**	**Phase V**
Planning: • Initiation of Organizational Exploration and Assessment; • Translate needs to purposes and goals; • Inventory and Assess resources	• Survey of Staff status and needs; • Individual and group conferences; • Secondary Assessment of community needs, resources; • Orientation and preliminary preparation of initial group for full participation	• Training and Commitment • Individual and Group (team) identification of organizational problems blocking increased productivity. • Plan(s) development for problem(s) resolution • Commitment to organizational change via group decision-making • Plans Assessment • Administrative Decision: — "Go or No" —	Implementation: (Impact) and follow-through • Project Teams receive reinforcing administrative support • Resources Committed • Targets Specified • Power Distribution to: • Reinforce Commitment • Assign Responsibility • Encourage Compliance • Begin to Institutionalize organizational renewal and development	• Rigorous Program Assessment • Program Modification prior to recycling • A provision to monitor, maintain and continue to evaluate committed program(s) until goal(s) achieved
Administrative Leadership	Administrative Leadership	Administrative Leadership	Administrative Leadership	Administrative Leadership

Figure 1. Organizational Redevelopment and Change (ORAC)

where the administrator finds himself in the role of an "equal" with his erstwhile subordinates, with each individual recognizing the other's separate responsibilities.

In addition there is some evidence that qualified group leaders can measurably enhance the quality of group work, not only through the projection of their non-superordinate image, but also through facilitating the effective presentation of materials and learning experiences. Chowdhry and Newcomb cited research which they interpreted to support their contention that "group understanding and knowledge, then, seems to be an important factor in the status that an individual may acquire in the group."[67]

While it cannot yet be called an assumption, the reader is asked to consider whether the day has already passed when school districts, especially those of small or medium size, can afford *not* to regularize the utilization of outside professional help, perhaps through establishing a continuing relationship with one or more institutions of higher learning.

School districts can hardly afford to regularly employ persons having the many types of special expertise needed from time to time in the pursuit of institutional excellence today. Besides, as one administrator put it, "A wise man is not one who tries to teach his own wife to drive."

Phase II — The Pre-Training (or Orientation) Phase — Individual conferences and small group interviews would be conducted by program leaders (or "trainers") who would use information previously obtained from unsigned completed questionnaires to govern their approach to preparing organizational members for maximum effort. Such questionnaires would, of course, be carefully tailored to the sociological dynamics of a particular institution and those serving it. Basically, it would be highly desirable to ascertain information concerning the expectations that individuals and groups hold for themselves and for each other as well as the organization. Particular attention would be given to employee perceptions related to morale, communication, decision making, power and authority, and current institutional policies, practices and procedures. Space does not permit an in-depth discussion of this important aspect of Phase II, nor would this be appropriate or even necessary. The reader is referred to the classic study, *Management and the Worker*,[68] for massive data regarding basic considerations holding importance for individuals serving primarily in subordinate capacities in organizations.

Especially important at this phase would be the development of a disposition for open communication, not only by subordinates, but admin-

---

67. Chowdhry and Newcomb, *The Study of Leadership*, p. 272.
68. F. J. Roethlisberger and W. J. Dickson, *Management and the Worker*.

istrators as well. All parties must somehow be made aware of the barriers to communication inherent in technical jargon and how the use of such language tends to inhibit the practice of asking questions.[69] It is important that one not be lulled into believing that each participant, including the administrator, has perceived and articulated the situation accurately.[70]

Indeed, it is probably crucial that a new awareness be established that all individuals perceive differently and that all parties be asked to continuously take this fact into account and openly deal with it in positive ways. It is difficult, but not impossible, for group members to be brought to an awareness and acceptance of the premise that the stated perceptions of others are neither necessarily inferior or superior to one's own; just different. "One must posit 'rational man' and try to determine how the situation interferes with his rationality."[71] Hence a major purpose of Phase II would be the preliminary preparation of group members, via reorientation to themselves and to the organization of which they are a part, for fuller participation in, and commitment to, organizational renewal and development.

Phase III – The Initial Group Training (or Workshop) Phase – Realistic considerations prompt the suggestion that this phase be limited to the number of contact hours estimated to be the minimum required to secure the actual involvement of participants in working groups, for the purposes of reaching a shared consensus regarding identified problems, and developing at least preliminary plans of action to attack and resolve them. New insights and skills growing out of the training received would, of course, not rule out the creative development of at least substantial preliminary plans to initiate modest, or even profound, change in the organization.

This phase—as in the case of the others—would be accompanied by rigorous evaluation, including the use of qualified observers. Appropriate modification and restructuring of this phase of the system would be mandated so as to provide (theoretically) an even more serviceable model for subsequent participants who are to follow the incumbents into Phase III.

It has been observed that organizations in their early or beginning stage, along with their leadership, tend to be shaped by the environment. However, organizations tend, as they become larger and more complex, to "outgrow the stage of strong leadership.[72] Instead, they tend to gravitate towards a stage of "management leadership"—where decision-making is diffused and where "the more of their decisions the members of a group

---

69. Halpin, *Administrative Theory in Education*, p. 11.
70. Thompson, *Educational Administration: Selected Readings*, p. 87.
71. *Ibid.*, p. 97.
72. Allen, *Administrative Theory*, pp. 30, 83.

can make, the greater their personal satisfaction and the greater the cohesiveness of the group.[73]

There is of course the very real question whether educational administration can be transformed to a type of leadership adequate to facilitate continuous renewal and development of the school organization, thus determining its predictable future. In addition to the certainties of law and tradition, there are the uncertainties of social change and values. Moreover, school organizations will probably always be hard put to match the practice of industry that ostensibly permits its workers a share in the more tangible benefits of increased productivity.

Phase IV — The Implementation (or Follow-Up) Phase — Basically this phase would include the giving of reinforcing support to established teams, as well as individuals, as they proceed to implement plans previously designed to ameliorate or resolve identified issues or problems. Specific procedures would be introduced at this stage, each designed to enhance or reinforce the participants' ability to more objectively perceive institutional problems, particularly those having their roots in human relations and communications. Individuals and working groups would receive additional training intended to further develop their personal skills which conceivably would then, when employed with success, help to further deepen the commitment by the individual to stay with his team until success attends their efforts. The assumption, "Nothing succeeds like successs" may well be more than theory.

Phase V — The Evaluation Phase — The individual and collective growth thus far achieved would (then) be recycled back into the system as the participants are asked to participate in a reassessment of the total program to date as a productive vehicle for initiating, implementing and sustaining organizational change. Other steps at this stage would be to elicit, evaluate and further develop additional recommendations for program modification and future action. And so the cycle and recycling continues. Administrative provision would of course be made to monitor, evaluate, and continue to maintain and support programs already committed. Any system of real worth must provide for dynamic feedback and semi-automatic mechanisms for intra-system clarification of differentiation of tasks, resources, responsibilities, targets and timing. In addition, a

---

73. *Ibid.*, p. 73 Note: This phase . . . really a small but important facet of a long-range program of institutionalized change . . . could conceivably achieve defensible results in any number of hours; however, a minimum of forty contact hours is suggested. Because of the sensitive nature of the enterprise, a minimum ratio of one qualified group leader to each ten participants is foreseen at the outset with a single workshop membership limitation of forty. However, the aim, clearly, would be to transfer responsibilities for group decision and direction to its members.

built-in capability for positive internal adjustment must be integrated into the total system. In a sense, the use of a systems approach is simply a carefully monitored and controlled, rationally determined set of procedures used to reach a number of dynamically related decisions. The aim is always to identify problems, however simple or complex, and develop promising ways (sub-systems) for solving them.

Perhaps one of the built-in values in the use of a systems approach is the insistence that analysis makes on the practitioner to clarify the assumptions which underlie his judgmental decisions. The carefully conducted analysis called for in the use of systems tends to call forth, and thereby reveal, the motivating factors rooted in administrative values already held.[74, 75]

In a process that depends so strongly on the application of intelligence to resolve problems and achieve pre-determined goals through program system design, implementation, evaluation, feedback and modification, care in analysis can be crucial.

The use of careful analysis can also be vastly important in identifying problems which are not amenable to system solution because of the limited availability of resources with which to attack them, or because of extremely narrow limits of flexibility inherent in the problem itself.

One thing, however, is certain: the administrator's "publics" are insisting on being involved in institutional decision making, including policy formulation. A systems approach to educational administration can point the way for their constructive, productive involvement without causing the machinery for institutional decisions to sputter, regularly mis-fire or grind to a complete halt. Indeed, the proposed model for a systems approach to educational administration through a process of organizational renewal and development, was designed to add new power and modern machinery to meet the need for greater institutional accountability and productivity, in addition to meeting the need for productive change in the educational establishment itself.

School administrators, whether they recognize it or not, employ theory as a basis for finding answers to questions, determining approaches to varied conditions or situations, and making decisions.[76] They make their decisions on the basis of what they "know," what they perceive they know, and what they think is appropriate or will work. Thus, "administrators make their decisions on the basis of generalizations and assumptions (hidden or explicit) and, in so doing, act in terms of a theory.[77]

74. *Educational Administration: Selected Readings*, ed. Hack and others, p. 377.
75. Saunders et al., *A Theory of Educational Leadership*, p. 32.
76. Halpin, *Administrative Theory in Education*, p. 21.
77. Coladarci and Getzels, *The Use of Theory in Educational Administration*, p. 4.

The use of theory by the administrator tends to give direction and consistency to his behavior and serves as a guide to new knowledge. At the same time theory provides the educational administrator, in common with his fellow administrators in other fields, with his own intellectual bases for his actions by causing him to identify principles and assumptions underlying his theory of educational administration (or leadership).[78] Stated in another way, when he operates from a carefully developed theoretical base, he consciously moves towards his decisions, using the resources available to him, including his own knowledge and intellectual powers, plus whatever additional inputs he can command or exploit. In so doing he strives to be realistic and objective, keeping in mind the goal(s) to be attained, and the possible alternatives open to him. When, instead, he operates from a position base which primarily consists of personal opinion, emotionally laden constraints or nonconscious bias, he probably can be depended upon to reach a goodly share of poor, inadequate decisions.[79] Such behavior and results will no longer suffice in today's world.

Theory, then, can help the administrator to better understand himself and, in so doing, help him to understand his constituency and how to better deal with it. The answers to the question on how to deal effectively with administrative problems will then, theoretically, be found within "good" theory. The problems may originate in the school community environment, but the genesis of the administrative approach to be employed in dealing with them must ultimately be found in the administrator himself. He can no longer ordain or direct that his will be obeyed; he must live and work with people, all the while enlisting their active support.

The writer was not concerned with theory for theory's sake. He has devoted three decades to the field of public education, nearly two-thirds of that time in top levels of school district administration. One aspect of the mission was to attempt to wed adminstrative theory to a systems application for purposes of institutionalizing continuous organizational change within the framework of a democratic philosophy.

That the mission was aimed at trying to understand what "is," in order to develop a system dependably capable of predicting behavioral results, does not gainsay the fact that the result is, in truth, only a modest beginning. Nevertheless, if such patterns of administrative performance do ultimately grow and develop into sophisticated practices capable of high reliability in predicting behavioral results, then the secrets of administrative competence cannot elude us for long thereafter.

78. Barnard, *The Functions of The Executive*, pp. 11-20.
79. *Ibid.*, p. 15.

It is for us, the learners, to be diligent though patient in our search for better theory. It was Halpin who claimed that "the development of theory in administration has been impeded by three substantive problems:

1. We haven't been clear on the meaning of theory.
2. We've tended to be pre-occupied with taxonomies and have confused these with theories and
3. We have not been sure of the precise domain of the theory we are seeking to devise.[80]

SELECTED REFERENCES

Allen, Louis A. "A Unified Theory of Administration." A paper presented at an Interdisciplinary Seminar on Administrative Theory, March 20-21, 1961 at the University of Texas: Austin, Texas. From *Administrative Theory*, pp. 21-36. Austin: The University of Texas, 1961.

Allport, Floyd H. *Theories of Perception and Concept of Structure*. New York: John Wiley and Sons, 1955.

Barnard, Chester I. *The Functions of the Executive*. Cambridge, Mass.: Harvard University Press, 1964. (Originally published in 1938).

Brodbeck, May. "Models Meaning, and Theories." *Symposium on Sociological Theory*, edited by Llewllyn Gross, pp. 373-403. Evanston, Ill.: Row, Peterson, 1959.

Buffington, Reed L., and Medsker, Leland. "Teachers and Parents Describe the Effective Principal's Behavior." *Administrator's Notebook* 4 September 1955, pp. 1-4, as reported by Wm. W. Savage. *Interpersonal and Group Relations in Educational Administration*, p. 136. Glenview, Illinois: Scott, Foresman and Co., 1968.

Carter, Lawnor F. "On Defining Leadership." in *The Study of Leadership*, edited by C. G. Browne, and Thomas S. Cohn, pp. 22-25. Danville, Illinois: The Interstate Printers and Publishers, Inc., 1958.

Casey, Thomas. "The Compatability of Free Choice with the Prediction of Human Behavior." *Rocky Mountain Science Journal*, October 1967, pp. 151-159.

Chowdhry, Kamla, and Newcomb, Theodore M. "The Relative Abilities of Leaders and Non-Leaders to Estimate Opinions of their Own Groups." in *The Study of Leadership*, edited by C. G. Browne, and Thomas S. Cohn, pp. 263-274. Danville, Illinois: The Interstate Printers and Publishers, Inc., 1958.

Coladarci, Arthur P., and Getzels, Jacob W. "The Nature of Theory—Practical Relationship." Reprinted in *Selected Readings in Educational Administration*, pp. 58-64. Boston: Allyn and Bacon, Inc., 1965.

80. Hall and Fagan, *General Systems—Yearbook of the Society for the Advancement of General Systems Theory*, p. 5.

Coladarci, Arthur P., and Getzels, Jacob W. *The Use of Theory in Educational Administration.* Monograph no. 5. School of Education, Stanford University, Stanford, California, 1955.

Cunningham, Luverne L. "Research in External Administration—What Do We Know?" in Roald Campbell and James Lipham, eds. *Administrative Theory as a Guide to Action.* Chicago: Midwest Administrator Center, University of Chicago, 1960.

Demeke, Howard J. *Guidelines for Evaluation: The School Principal—Seven Areas of Competence.* Tempe, Arizona: University Bookstore, Arizona State University, 1971.

Educational Administration: Selected Readings, W. J. Hack, J. A. Ramseyer, W. J. Gephart, and J. B. Heck, editors. Boston: Allyn and Bacon, Inc., 1965.

Feigl, Herbert. "Principles and Procedures of Theory Construction in Psychology." in *Current Trends in Psychological Theory,* p. 182. Pittsburgh: University of Pittsburgh Press, 1951.

Feroleto, Leonard. "An Analysis of Educational Leadership in a Democratic Setting." Doctor's dissertation, Arizona State University, 1969.

Getzels, Jacob W.; Lipham, James M.; and Campbell, Roald. *Educational Administration as a Social Process.* New York: Harper and Row, 1968.

Gibb, Cecil A. "The Principles and Traits of Leadership." in *The Study of Leadership,* edited by C. G. Browne, and Thomas S. Cohn, pp. 67-75. Danville, Illinois: The Interstate Printers and Publishers, Inc., 1958.

Goldhammer, Keith. *Issues and Strategies in the Public Acceptance of Change.* Eugene: University of Oregon, Center for Advanced Study of Educational Administration, 18 November 1965.

Griffiths, Daniel E. "Administration as Decision Making." in *Administrative Theory in Education,* pp. 119-149. Chicago: The Midwest Administrative Center, University of Chicago, 1958.

Griffiths, Daniel E. *Administrative Theory.* New York: Appleton-Century-Crofts, Inc., 1959.

Griffiths, Daniel E. "Administrative Performance and System Theory." One of several papers presented in *Administrative Theory.* Austin: The University of Texas, 1961.

Griffiths, Daniel E. *Human Relations in School Administration.* New York: Appleton-Century-Crofts, Inc., 1956.

Hall, A. D., and Fagan, R. E. *General Systems—Yearbook of the Society for the Advancement of General Systems Theory,* edited by L. V. Bertalauffy and A. Rapport. Ann Arbor: Braun-Brumfield, 1956.

Halpin, Andrew W. *Administrative Theory in Education.* Chicago: The Midwest Administrative Center, University of Chicago, 1959.

Halpin, Andrew. *Theory and Research in Administration.* New York: The Macmillan Company, 1966.

Halpin, Andrew. *The Leadership Behavior of School Superintendents.* Columbus, Ohio: College of Education, Ohio State University, 1956.

Hemphill, John K.; Griffiths, Daniel; and Fredericksen, Norman. *Administrative Performance and Personality*, p. 339. New York: Bureau of Publications, Teachers College, Columbia University, 1962.

Kaufman, Roger A. "Accountability, A Systems Approach and the Quantitative Improvement of Education—An Attempted Integration." *Educational Technology*, January 1971, pp. 21-25.

Learned, Edmund P., and Sproat, Audrey T. *Organization Theory and Policy*. Homewood, Illinois: Richard D. Irwin, Inc., 1966.

Lipham, James. "Leadership and Administration." *Yearbook of the National Society of the Study of Education, 1964,* Chapter 6. University of Chicago Press, 1964.

Meadows, Paul. "Models, Systems and Science." *American Sociological Review* 22 (1 February 1957).

Parsons, Talcott. "Some Ingredients of a General Theory of Formal Organization." in *Administrative Theory in Education*, edited by Andrew Halpin, pp. 40-72. New York: The Macmillan Company, 1958.

Roethlisberger, F. J. "The Administrator's Skill: Communication." Reprinted from the *Harvard Business Review*, November-December 31, pp. 55-62.

Roethlisberger, F. J., and Dickson, W. J. *Management and the Worker*. Harvard University Press, 1946.

Saunders, Robert L.; Phillips, Ray C.; and Johnson, Harold J. *A Theory of Educational Leadership*. Columbus, Ohio: Charles E. Merrill Books, Inc., 1966.

Savage, William W. *Interpersonal and Group Relations in Educational Administration*. Glenview, Illinois: Scott, Foresman and Company, 1968.

Saxe, Richard W. "The Principal and Theory." in *Perspectives on the Changing Role of the Principal*. Springfield, Illinois: Charles C. Thomas Publisher, 1968.

Simon, Herbert. *Administrative Behavior*. New York: The Macmillan Company, 1956.

Stogdill, Ralph M., and others. "A Factorial Study of Human Behavior." in *The Study of Leadership*, edited by C. G. Browne, and Thomas S. Cohn, pp. 357-366. Danville, Illinois: The Interstate Printers and Publishers, Inc., 1958.

Thompson, James D. "Modern Approaches to Theory in Administration." in *Educational Administration: Selected Readings*, edited by W. J. Hack, J. A. Ramseyer, W. J. Gephart, and J. B. Heck, pp. 82-100. Boston: Allyn and Bacon, Inc., 1965.

Walton, John. "The Theoretical Study of Educational Administration." *Harvard Educational Review*, 1955.

Weschler, I. R.; Klemes, M. A.; and Shepard, C. "A New Focus on Executive Training." in *The Study of Leadership*, edited by C. G. Browne, and Thomas S. Cohn Publishers, Inc., 1958.

3

JAMES M. THRASHER
LYAL E. HOLDER

The Administrative Organization for Implementation of Educational Purposes

Any organization exists for a purpose. It is not an end in itself, it is a social artifact created by man to accomplish certain broad or specific purposes. Its success cannot be measured by the mere act of being able to stay alive or to function or to perpetuate itself, but is measured in terms of its ability to perform its purpose. The administrative organization in education is part of a larger organization. Success is measured by the contribution that it makes to the total educational enterprise. Scholars often find that a small effective operating group becomes more and more concerned with the problems of maintaining the administrative superstructure of the organization, rather than getting outside of that administrative structure and looking at capacity to accomplish the objectives or purpose of the administrative organization.

The administrative organization in the educational setting has a very specific purpose. It is to create, maintain, nurture and promote a climate in which the learners can interact with the total educational environment to grow in their capacity as human beings, to learn tool skills, to explore and find self realization, and to become addicted to life-long learning.

The administrative structure in the educational enterprise is an absolute necessity. Without positive and effective coordination, logistical support, and central decision making, the school district, or the educational agency, would be like a raft upon the high seas, free to move in whichever direction the winds or the waves happen to be blowing at a given time. If the administrative structure is geared to accomplishing its purpose of supporting an environment and climate for learning to take place, it has a positive potential for education. If the administrators in the organizational framework become overly obsessed with examining and treating the problems of keeping the administrative organization moving, they have little time for the support of the entire educational enterprise. Under such conditions the superstructure can become a negative potential.

It is a mistake to judge the effectiveness of the administrative organization in education by the size, i.e., the number of people engaged in the central office or other administrative functions of a school district. Rather, the degree of success should be looked at in terms of how well the administrative organization truly supports the main mission of the schools—instruction—in creating a climate for learning.

EDUCATIONAL ADMINISTRATION—MAJOR FUNCTIONS

The existence of the educational administration organization rests upon six major functions to be carried out or served. These are: decision making, communication, planning, implementation, evaluation, and accountability.

The Decision Process

Lasswell has identified the decision process—seven categories of functional analysis.[1]

The act of gathering information, making predictions based upon that information, and planning based upon digestion of information, has been called the *intelligence function.*

The *recommending category* of functional analysis is described as handling of pressures for change in the promotional activities of pressure groups or organizations that seek to alter the course of their own, or other organizations.

The *category of prescribing* involves establishing rules on a more or less tentative basis for trial and also involves designing new policies or programs.

The *invoking category* has been described as the act of provisionally trying out suggestions or plans that have been promoted and includes policing or supervisory aspects for this category of functional analysis.

The *application category,* or final act of alloting money, space, or personnel to carry out an activity, carries with it the responsibility for implementing rules or verifying that prescriptions have been followed.

Appraisal activity is the sixth category. The major concern is that prescriptions are carried out in a congenial and cooperative way and in an

---

1. Harold D. Lasswell, *The Decision Process: Seven Categories of Functional Analysis* (College Park, Maryland, 1965), p. 2.

economically efficient manner. The appraisal activity also includes an assessment of success or failure of the prescriptions or policies.

The final category described in the decision process is that of *terminating activities*. It would include the suspending of programs, making amendments to policies, and closing out or phasing out activities of the organization, if and when they should be phased out.

Communication

Individual members of an organization more actively support those decisions which affect them when they have a voice in development of those decisions. This involvement requires the communication process.

Communication is the art of selecting, sending, receiving and interpreting arbitrary symbols which represent specific meanings. Shakespeare said it this way, "A rose by any other name would smell as sweet." A rose would still be what it is regardless of the arbitrary symbol selected to represent it. It is the meaning that is to be communicated, not the symbol.

Communication efforts intended to solicit the active support of members of an organization must meet the test of four basic criteria.

First, they must be capable of being understood. The symbols selected must cue up meaning for the receiver which is as congruent with that held by the sender as human perception will allow. The more this is left to chance, the greater the risk of the communication not being understood and, therefore, not accepted.

The second basic criterion to be met is that the communication must solicit support which the receiver is physically and mentally capable of giving.

The third criterion to test communications is that they must be seen by the receiver as evidencing congruence with organizational goals, as he perceives them.

The fourth criterion is that communications must be judged by the receiver as being in harmony with his personal goals.[2]

Effective intra-organization communication suggests that the members of that organization know what the goals of the organization are and see them as their goals for the organization as well as goals that they can subscribe to that are in pursuit of their own personal life.

The administrative organization of a formal institution must maximize intra-organizational communication. That organization must insure that the communication process is a two-way vehicle. For example, an administrative organization should be sensitive to acceptance or nonacceptance of a communication dealing with organizational purposes.

---

2. Chester I. Barnard, *The Functions of the Executive* (Cambridge, Massachusetts: Harvard University Press, 1968).

Non-acceptance is a denial of authority or of support from a member or members toward accomplishment of the purpose of the communication. Such a condition requires immediate investigation to find out in greater detail the reasons for the non-acceptance. This action may be only a symptom, not the cause, of the non-support. The investigation must be carried out in a personalized way so that the organization is seen by the individual as being aware of his needs. This requires a keen sensitivity to the intrapersonal relationships that exist between and among all members of the organization as well as the ability to marshall the resources within the institution to help develop communication congruence. Such a condition may require modification of the organization as much or more than that of an individual or group of members. The first step requires a look at the administrative organization. The larger the administrative organization the more internal inertia it contains and the greater the force required to change its course.

Administrative organizations that emphasize responsible and comprehensible patterns of two-way communication, sensitive to the self-image and self-worth of each member, will be more effective and efficient in soliciting, obtaining, and utilizing the contributions of their members in determining and accomplishing organizational goals.

Planning

Barnard suggested in *The Functions of the Executive* that there are three essential elements of an organization. These are communication, willingness to serve, and having a common purpose.[3] It is this third element, common purpose, in which the function of planning creates the so-called common purpose of the organization, and develops potential strategies for accomplishing the organizational goals.

School administrators have often made the mistake of carrying out planning at the top echelons of the administrative organizational plane and handing down well-formulated plans to the people at the middle management, or at the school site level, to carry out these plans. This handing down of the prescriptive plan without direct involvement of those who are responsible for implementing those plans almost always results in mediocre performance, inefficiency, or outright rejection of the plans. By the time these plans reach the practitioner in the classroom they have been transmitted and interpreted a minimum of three times. The personnel responsible for the day-to-day implementation of plans may have limited knowledge of the original needs assessment that brought about the plans,

3. *Ibid.*, p. 82.

the points for re-cycling, or the evaluative criteria to be applied in assessing the value of the chosen activities or direction.

Effective planning must be accomplished by and through the administrative organization in such a manner as to involve personnel who are to be affected by planning and to give such personnel a place in formulating the plans which they are expected to be a part of in the implementation stage.

One can almost measure the temperature of an organization in the vibrance, or lack of vibrance, with which teachers and middle management administrators discuss their part in the total organizational scheme. Halpin has described this phenomena of feel as the organizational climate of schools.

> Anyone who visits more than a few schools notes quickly how schools differ from each other in their feel. In one school the teachers and the principal are zestful and exude confidence in what they are doing. They find pleasure in working with each other; this pleasure is transmitted to the students who thus are given at least a fighting chance to discover that school can be a happy experience. In a second school the brooding discontent of the teachers is palpable; the principal tries to hide his incompetence and his lack of a sense of direction behind a cloak of authority, and yet he wears this cloak poorly because the attitude he displays to others vacillates randomly between the obsequious and the officious. And the psychological sickness of such a faculty spills over on the students who, in their own frustration, feed back to the teachers a mood of despair. A third school is marked by neither joy nor despair, but by hollow ritual. Here one gets the feeling of watching an elaborate charade in which teachers, principal, and students alike are acting out parts. The acting is smooth, even glib, but it appears to have little meaning for the participants; in a strange way the show just doesn't seem to be "for real." And so, too, as one moves to other schools, one finds that each appears to have a "personality" of its own. It is this "personality" that we describe here as the "Organizational Climate" of the school. Analogously, personality is to the individual what Organizational Climate is to the organization.[4]

The advent of the so-called systems approach to planning has spawned a new breed of person within the organizational structure of the educational enterprise. He may be known as the system(s) analyst or simply as the educational planner. The systematic (logic) scheme in planning has existed in some form within educational organization for some time. It has not always been used or understood by as many administrators as should have been utilizing the technique. The administrative organization must invoke the systematic approach for attacking the educational

---

4. Andrew W. Halpin, *Theory and Research in Administration* (New York: The Macmillan Company, 1966), p. 131.

problems of our time. However, the systems approach to planning must never be allowed to become an end in itself. It is a process by which the organization can move more effectively toward accomplishing the central mission of the educational organization.

The new breed within the administrative organization can play an important role in planning. His expertise in assisting to guide the planning process can be most helpful. A real danger exists if it is assumed that such a person can take over and do the planning necessary to effect movement. Effective planning cannot be done in isolation from the total organization and then imposed. It is a team operation. The trained and specialized planner is one member of that team.

IMPLEMENTATION

Administrative organizations are causing agents. They exist to guide, to involve, and to cause something to happen that moves the total organization toward fulfilling its responsibilities or goals.

Implementation, the act of installing or getting something underway, does not happen by accident or osmosis. It must be consciously planned as carefully as we plan the format of a curriculum change. Until the act of implementation takes place the administrative organization has not really moved the organization any closer to realizing goals than it was before new planning took place. An appropriate analogy might be that the aerospace industry expends resources on designing and building a prototype jumbo airliner and continues to keep it in the hangar; never working it out to the end of the runway for actual flight.

Administrators within the administrative organization must keep in mind that until the implementation stage is reached, the work of the organization in moving toward fulfillment of its mission or purpose, has not really occurred. Unfortunately, some administrators have found it most difficult to take Lasswell's fifth step in the "Decision Process"—the application function of allotting money, space, and personnel to carry out a planned program. Failure to give attention to the area of installation—implementation—can be costly in eroding away the confidence that the total organization holds for the administrative organization of the school district and can destroy public confidence in educational change. It may be that the reluctance on the part of the public to face and accept change in education is more related to the way we have implemented new programs rather than any inherent weakness in the change itself.

Timing of the implementation phase is also a critical issue. Teachers are often heard to lament that they spend all the time in planning, or understanding the problem, instead of getting the work done, or applying a solution to the problem. This undoubtedly is a gross exaggeration, but does point up the fact that the administrative organization must function to clearly delineate the planning phase and the implementation phase. This should not suggest that continuous planning and potential re-cycling of activities cannot go on simultaneously with application. The wise administrator must know when to stop talking about what is going to be done, and begin to do it.

Evaluation

The administrative organization for the implementation of educational purposes must be able to perform, or cause to be performed, the function of evaluation. Evaluation may take many forms, but it will need to focus on both the performance of the total educational enterprise which the administrative organization serves, and the internal administrative structure within the administrative organization.

Social scientists have been slower in arriving at more precise evaluation techniques than have some disciplines. This can easily be explained because the social scientist works in the human context, and precise measurement of events and outcomes may not be so readily apparent as measurements of things. Since the last half of the 1960's there has been a concerted effort to pull together the known evaluation processes in the field of education and to create new models for evaluation. It is not the intent here to endorse or advocate one evaluation system or model. It is the intent to point up, as strongly as possible, that the administrative organization that is going to be effective in the implementation of educational programs must be aware of and use appropriate evaluation techniques. The framework for evaluation is set by the administrative organization during the planning stage. In "days of yore" it might have passed for the administrator to say after a program or practice had been in operation for a period of time, "Now it is time we establish how we are going to evaluate our new reading program." That is not the case today. Even the administrator nominally informed knows how imperative it is to establish goals and appropriate evaluation devices to monitor the programs of the implementation schema, as well as decide at the outset how programs within the effect are to be judged. The same administrator also knows that to be most effective, all of the internal and external publics concerned with the institutional purposes should be involved directly, or by accepted and respected representation, in determining the evaluation criteria, methods of assessment, and interpretation and dissemination of the results.

The other facet of evaluation that the administrative organization must deal with is how to determine its internal effectiveness, as well as, how the various parts of that organization support the work of the whole institution in attaining its primary goal. It is increasingly apparent, as public clamor has risen over costs of education and as teacher organizations have raised questions about the number and duties of administrative staff personnel, that evaluation of the administrative organization and the components that make it up must have clear cut performance objectives that are clearly established. Evaluation of the administrative organization should be a continuous process with opportunities for re-cycling planning during the progress of work as well as periodic checks on how well objectives are being met.

Accountability

Accountability of the administrative organization of schools is an accepted fact in the 1970's. The only questions are who is the organization responsible to and for what?

Local school boards could answer the first question quite easily. The administrative organization is responsible to them. It should be held accountable for the effectiveness of institutional members, and resource persons drawn upon from outside the institution, in fulfilling their role expectations; effectiveness of the support system for helping them to become effective; and the decision to recommend removal from membership those who do not, or cannot, become effective. In other words, the administrative organization is accountable to the board of education for the effectiveness of its leadership in insuring that all those involved in institutional tasks, especially those within the administrative organization, are accountable for fulfilling their roles.

The local school board hires the superintendent and expects him to mold the administrative cadre into an organization that is both effective and efficient. Effectiveness is used here to denote the qualities of being able to guide, support, and nurture the educational enterprise in attaining its primary goal, that of instruction. Efficiency alludes to the cost of utilizing institutional resources in accomplishing its work. An administrator has only three resources with which to work. These are *Time, People,* and *Money*. The allocation of these three basic resources in proper proportion at the right moment in history is probably one of the greatest indices of successful administration. Boards of education seldom realize that bringing in a new superintendent without giving him latitude in organizing, reorganizing, or reassigning people within the administrative organization is tantamount to asking him to be accountable when the resources he is to use have been severely curtailed. Recent limited advocacy of an administrative team approach to filling the top position in a school district may have some validity if newly employed chief administrators are not given wide latitude in managing the personnel resources that exist within the administrative organization at the time they take up a new assignment. Like the relief pitcher, he should be accountable only for the runners that he put on base.

Teacher organizations could also respond readily to the question of who the administrative organization is accountable to and for what. The answer would be somewhat different than the view held by board members. Out of the drive for more autonomy for the professional association, and the teacher's union, there has emerged a view of administration as fulfilling only a limited managerial role. The educational leadership function would be supplied solely from the teacher corps via their association. Accountability of the administrative organization would be measured in terms of support for the programs of the teacher's association.

The general public has slowly become aware of the need for the ordinary citizen to take a personal interest in the great American dream—an education for all, at least through the 12th grade. The cost in dollars of fueling the machinery to provide that educational opportunity has caused taxpayers to raise the question of efficiency in educational administrative organizations. The drive for equality of educational opportunity, or results, without regard to race, religion, ethnic background, or location of residence has created additional dimensions of concern in the public mind about the effectiveness of the educational administrative organization to deliver the services and product quality expected. The publicity surrounding "accountability" as it pertains to governmental functions, particularly in the defense industry, has led to an increased interest on the part of the lay community in invoking some additional qualitative measures in assessing the effectiveness of the educational effort. This interest has even caused the question to be raised concerning alternative routes to accomplishing the instructional missions. "The most dramatic current attempt to assume responsibility for education is the involvement of private instructional firms which are willing to be paid on the basis of student performance."[5] Performance contracting, regardless of success or failure, is a symptom of the times indicating the great desire of the general public to hold the administrative organization of their schools accountable for the product they produce.

THE NATURE OF EDUCATIONAL ADMINISTRATIVE ORGANIZATIONS

The complex organization of schools and the administrative organization associated with schools is a development of the last thirty to forty

---

5. Reed Martin and Charles Blaschke, "Contracting For Educational Reform," *Phi Delta Kappan* (March 1971), pp. 403-405.

years. It might be argued that some of the larger cities have had rather extensive administrative structures for a longer period of time. However, the majority of school districts were small and most administrative functions were accomplished by one individual or in some cases by personnel who were part teacher and part administrator.

As the scientific management movement grew in business and industry in the United States, it was inevitable that it would have some influence upon school administration. The pace of American life quickened and more and more expectations were held for education. The advent of increased numbers of people attending school and the growing diversity of program dictated the growth of an administrative structure to guide the increased activity.

The so-called "line and staff" organization chart has become a familiar part of the working arrangements within school districts. The staff positions are usually described as advisory or as having no direct authority for the operations phase of school management. The line positions are characterized as having direct operating responsibilities with each line officer reporting along the organization chart to the next higher line officer. In actual practice in today's educational administrative organization it is difficult to find that clean line which differentiates the line and staff function. Another characteristic of the line-staff organizational framework is that the moment it progresses beyond ten or twelve people it consumes time in large quantities to keep the system functioning. It is necessary to safeguard each line position in its function. Therefore, a request for change or movement must traverse the vagaries of the organizational chart. This can be a time-consuming process. It also has a tendency to insulate the chief administrative officer from hearing or interacting with the middle-management team. His communication, unless he initiates some alternative routes for it, can be limited to the immediate personnel around him such as the associate or assistant superintendent.

The weaknesses and slowness of this traditional organizational structure of administration to meet the needs of education in our time is causing the serious students of school administration to cast about for alternatives that might be more effective.

The move to decentralize the administrative structure is one such effort. It might be described as a sort of half-step in seeking an alternative to the rigid organizational chart scheme. In most cases the decentralization has consisted of breaking the line-staff arrangement into several smaller units, but with a tie to the central administrative unit.

Another alternative that has been suggested, and alluded to earlier in this chapter, is the administrative team approach to organizational structure. The team would consist of a group of trained individuals who would contract with a local school board to provide the central administrative organization for the district. The top echelon administration would all be the province of the team. They would relate directly to the site administration in the individual schools. The size and complexity of the team would vary with the needs of the district. When board confidence, or public support for the administration waned, the team would be replaced instead of individual administrators being terminated. If boards of education truly want to bring about change in the school organization and want to hold the chief administrator responsible for the conduct of the business of the educational enterprise they may want to examine closely the potential of the central administrative team approach.

The demands of the times and the needs of education to adapt and change may indicate that a "radical" new approach to administrative organization is needed. At this time there is a dearth of models that truly present alternative routes for providing administrative support and leadership. One model that suggests adaptability to meet changes rapidly and to constantly undergo reorganization for mission accomplishment is the "ellipse organization." A schematic of the ellipse is presented in Figure 1. The ellipse model is based upon several rather fundamental assumptions. One of those assumptions is that the needs of the educational enterprise are constantly shifting and the expertise needed by personnel engaged in planning and implementing programs must be readily adaptable to those changing needs. Another assumption is that a school district cannot afford to employ and expect to retain for years the top people in the specialized fields impinging upon the work of education, but that it is possible to identify top people and engage them on a sustaining basis for a long enough period of time to plan and implement educational programs. It is also an assumption that more expertise does exist in the teaching corps of the school district than has been historically tapped. The aim of the ellipse model is to tap that resource for a period of time. The involvement of teachers for specific mission assignments may pay rich dividends in capturing their ideas and views as well as render the organization more vibrant because of the involvement. The last assumption upon which the ellipse model was created is that there is no way to know far in advance what the very specialized needs of the administrative organization may be. If all of the financial resources are committed, there may be no way to tap the specialized knowledge needed. There needs to be a planned place for the assignment of the top experts who can be located to work within the administrative structure for short periods of time. Such experts would most likely be so highly specialized, and sought after, that there would be no thought of attempting to attach them to the permanent administrative organization. Their specialty may be needed only for short

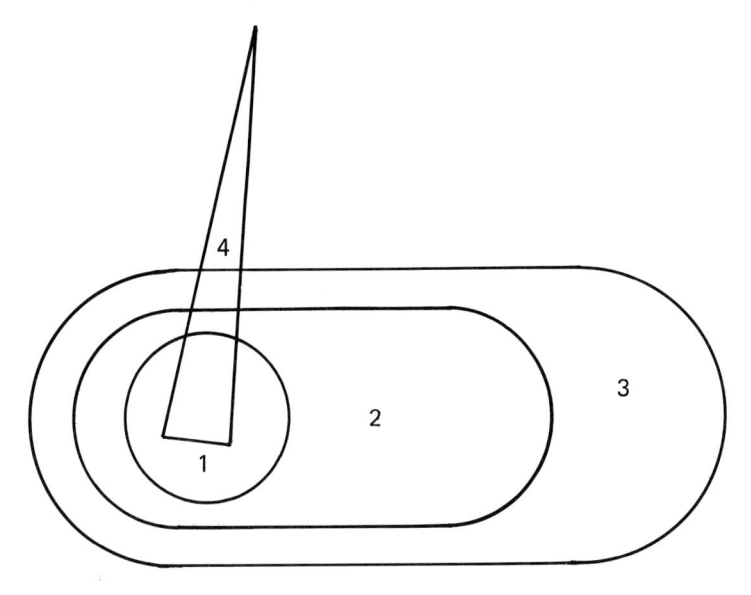

1. Central minimum core of administrative personnel to provide continuity in planning and accountability.
2. Specialized personnel both from within and from without the school district who can be organized into task forces for mission assignment. (They are persons who have continuing contact and know the personnel and school district well.)
3. The instructional corps (teachers) from which personnel can be drawn for assignments for a period of time to task forces.
4. Outside special experts who are employed to accomplish a single task or mission and may be working with personnel in any assigned level.[6]

Figure 1

periods of time. The ellipse organization has been described here, not as *the* answer to administrative organization, but as one possible alternative to remaining chained to the traditional line-staff type structure.

THE FUTURE OF EDUCATIONAL ADMINISTRATIVE ORGANIZATIONS

There are no crystal balls to which one can turn to ascertain exactly what the future holds for the metamorphosis of the educational administrative structure. One thing can be predicted with certainty. The organization of administrative services in the educational enterprise will change. The needs of the primary educational organization which is served by the

6. James M. Thrasher, The Institute For Educational Management, United States International University, San Diego, California.

administrative structure change and shift in emphasis. It follows that administration must change to meet these shifts in needs or endanger the entire organization. In addition to the inherent needs theory which is reshaping educational administration, at least three organized forces are at work which greatly influence the shape and structure of administration. These forces are the general public, the teacher organizations, and school boards.

Public interest in education has been inherent in the very nature of the American educational system since its inception. However, until the 1950's, the vast majority were willing to assert the view that professional educators and administrators should determine how the schools should be operated. The advent of the "critics" of American education such as Bestor has focused an increasingly inquiring eye upon the conduct of education.[7] The general public not only wants to have more information as to how things are being done, but they are strongly requesting a voice in the decision making process. The administrative organization must find new ways of informing people and providing avenues for public participative management of the schools. In the final analysis the schools do belong to the people. They give every evidence of making themselves heard.

The drive of the professional association of teachers to become more involved in the administration of the school district has had a dramatic effect in a few short years. Associations, through the negotiating process, are influencing the number, the functions, and the salaries of personnel in the administrative organization. There is a view in at least some quarters of the association that teacher's organizations themselves should become the administrators, at least at the school site level. All instructional and administrative decisions would then be in the hands of the organized teacher groups. This view is rather radical when examined in the context of the times and the extreme needs of education. However, it would be fallacious to assume that the influence of the teacher organization was not causing some worthwhile soul searching and the changing of administrative procedures in almost every school district in the United States.

School boards are receiving increased attention through the media. As costs of education mount, a certain segment of the population raises questions. As the problems of desegregation and local control are debated, another sector is heard from. The quality of the educational product is now being widely debated. These factors both singly and together are causing boards of education to react differently than they did ten years ago. There is a great need for board members to have more information

---

7. Arthur E. Bestor, *Restoration of Learning, A Program for Redeeming the Unfulfilled Promise of American Education* (New York: Knopf Publishers, 1956).

and to be able to conduct the affairs of policy decision making under the critical eye of an interested constituency. The political nature of boardmanship makes it imperative that the board member feel secure that he is getting adequate information in a timely fashion and that he is being dealt with in a manner consistent with that accorded all other board members.

The intensity of the problems facing most school boards creates in the operational relations between the administrative organization and the board real need for cooperation, for adequate information flow, and for mutual trust. Isolation of a board from the real world of running the school or interference of the board in administrative functions are both unrealistic if the administrative organization is to be truly effective and efficient. It behooves the chief administrator to find new or improved patterns of administrative structure that will fully meet the needs of boards of education in their important and sometimes stressful position.

SUMMARY

The complexity of the problems impinging upon the educational enterprise and the technological capability available to solve these problems requires administrators and administrative organizations that can meet the challenges of the times. If you please, new wine in new bottles.

The administrative expertise needed is that of leadership, the ability to communicate and work with all of the publics responsible for and to the educational enterprise more than the specific tool skills required to solve current and unforeseen problems related to instruction, business and personnel. The challenge is to capitalize upon the commonalities among the diversities of these publics for determining the purposes of the organization(s) at all levels in terms of the product to be produced, the best means of accomplishing these purposes, monitoring the process, evaluating its results and bringing about change, if needed.

The traditional self-contained line-staff administrative organization with all of the specific expertise deemed necessary to meet every unforeseen exigency is only one alternative. The administrative team and ellipse organizations are two others.

While the self-contained line-staff organization moves like the stolid, flat-footed slugger to solve its problems, the administrative team moves with the stealth of a good hitter and excellent counter puncher with adequate footwork. The ellipse organization adds to this superior footwork and the hidden punch. It provides the flexible, catalytic climate for ferreting out and welding together the precise expertise needed, whether within or from without the organization, for only the amount of time required. It better holds the promise of helping the diverse groups that have direct concern for the educational enterprise to capitalize upon their commonalities and minimize their differences because it must look outside of itself for most of the specific tool skills needed to solve organizational problems. These constant searching and unifying efforts can legislate against ponderous self-sufficiency, the administrative organization disease.

Selected References

Atkins, Thurston A. "It's Time for a Change—Or Is it?" *The National Elementary Principal* 48 (1969).

Barnard, Chester I. *The Functions of the Executive.* Cambridge, Massachusetts: Harvard University Press, 1968.

Bestor, Arthur E. *Restoration of Learning, A Program for Redeeming The Unfulfilled Promise of American Education.* New York: Knopf Publishers, 1956.

Bush, William T. "What Administrators Do to Improve Instruction." *Phi Delta Kappan,* November 1959.

Erlandson, David A., and House, Ernest R. "Theory and Practice: Why Nothing Seems to Work." *Bulletin of the National Association of Secondary School Principals,* Vol. 55, No. 354 (April 1971), pp. 69-75.

Griffiths, Daniel E. *Administrative Theory.* New York: Appleton-Century-Crofts, Inc., 1959.

Griffiths, Daniel E., and others. *Organizing Schools for Effective Education.* Danville: The Interstate Printers and Publishers, Inc., 1962.

Halpin, Andrew W. *Theory and Research in Administration.* New York: The Macmillan Company, 1966.

Kimbrough, Ralph B. *Administering Elementary Schools.* New York: The Macmillan Company, 1968.

Lasswell, Harold D. *The Decision Process: Seven Categories of Functional Analysis.* College Park, Maryland, 1956.

Levine, Daniel U. "The Principalship in Schools That Are Coming Apart." *Bulletin of the National Association of Secondary School Principals,* Vol. 54, No. 349 (Nov. 1970), pp. 24-39.

Martin, Reed, and Blaschke, Charles. "Contracting for Educational Reform." *Phi Delta Kappan,* March 1971.

Miles, Matthew B. *Innovation in Education.* New York: Bureau of Publications, 1964.

Morris, Russell J. "The Administrator and Public Relations." *National Association of Secondary School Principals,* 44 (1969).

Neagley, Ross L., and Evans, N. Dean. *Handbook for Effective Supervision of Instruction.* Englewood Cliffs: Prentice-Hall, Inc., 1970.

Parsons, Talcott. *Structure and Process in Modern Society.* Glencoe, Illinois: The Free Press, 1960.

Wayson, William W. "A New Kind of Principal." *The Education Digest* 36 May 1970).
Wiles, Kimball. *Supervision for Better Schools.* Englewood Cliffs: Prentice-Hall, Inc., 1967.
Wogamon, Thomas D. "The Making of a Principal." *Bulletin of the National Association of Secondary School Principals,* Vol. 55, No. 354 (April 1971), pp. 76-79.

CHARLES ATWELL
DEWEY A. ADAMS
WAYNE M. WORNER

The Administrative Team Concept

What will the schools of the future be like? Will the role and tasks of the school administrator of the next decade differ significantly from today's administrative role? Will the same administrative strategies which are effective in the management of schools in 1972 prove equally effective in 1982?

While no one can answer these questions with confidence, the nature of the educational enterprise and the society in which it functions permit some logical speculation about these questions. For example, in an exciting recent publication entitled *Educational Futurism 1985*, Robert Ohm speculates:

School systems will become exceedingly complex. Specialization in position and role will continue with a significant increase in the number of specializations required to form functional administrative teams. Many, if not most of these specializations will require specific and rather narrow training with an emphasis on requisite skills. The traditional triology of buildings, buses, and bonds together with such new areas of administrative concern as computers, community action, and collective bargaining requires staff with intensive but specialized training. *It is recognized that the maintenance of the system will depend more and more on trained specialists organized in administrative teams, headed by a "comprehensivist" administrator.*[1] (emphasis mine)

The management of the educational enterprise has never been a simple task. The role of the school administrator, at all levels, has always been

---

1. Robert E. Ohm, "The School Administrator in 1985," in *Educational Futurism 1985*, Chapter 6, pp. 93-108 (Berkeley, California: McCutchan Publishing Corporation, 1971).

demanding and complex. No one would argue that the tasks of the college president, the school district superintendent, and the principal of two decades ago—or longer—were simple or uncomplicated. The fact remains, however, that today's society is infinitely more complex than yesterday's. The social institutions through which such a society's goals and objectives are achieved logically become, themselves, more complex.

It does not require scholarly pronouncements nor sophisticated research findings—although there is no dearth of these—to reason that increasingly complex institutions become increasingly more difficult to administer or manage. Most educational practitioners would give willing testimony to such a phenomenon. One has only to attend any gathering of school administrators to hear of the trials and tribulations of the educational manager—if not from the speaker's platform, certainly in the informal discussions which inevitably ensue.

In addition to the perennial problems which have faced school administrators since the beginning of organized education, a new and demanding series of problems has arisen in recent years.

Today's administrator must be prepared to deal with communities which have begun to realize that the schools are "their" schools and who no longer blindly accept educational practices at face value. Coupled with this awareness is a restiveness and growing dissatisfaction with many aspects of public education. Recent publications by Silberman,[2] Gross and Gross,[3] Glasser,[4] and Fantini and Young[5] have received popular acclaim and are probably symptomatic of this public concern over education.

Student unrest and resultant disruptions are no longer unexpected on our college campuses and are occurring with increasing frequency in our secondary schools.

Teaching faculty in today's schools are turning with growing frequency to strong teacher organization and unions concerning such problems as teacher welfare. Collective bargaining agreements frequently leave the principal in a position roughly analogous to that of Edward E. Hale's "Man Without a Country." Teacher "militancy," despite the negative connotations of the term, is no longer a myth.

The financing of public education, always a concern of the educational leader, can, without fear of contradiction, be termed a critical prob-

2. Charles E. Silberman, *Crisis in the Classroom: The Remaking of American Education* (New York: Random House, 1970).
3. Ronald Gross and Beatrice Gross, eds., *Radical School Reform* (New York: Simon and Schuster, 1969).
4. William Glasser, *Schools Without Failure* (New York: Harper & Row, 1969).
5. Mario D. Fantini and Milton A. Young, *Designing Education for Tomorrow's Cities* (New York: Holt, Rinehart and Winston, 1970).

lem in today's schools. Due in part to the element of public dissatisfaction mentioned above, and certainly partly a result of competing demands for a larger share of the public tax dollar, school districts throughout the country find themselves underfinanced. School millage elections and bond issues for capital outlay are being voted down at a rate unequalled in our history. Recent court decisions indicate that the entire financial support structure of public education may be drastically altered in the immediate future.

Forced desegregation of schools, while long overdue on legal and moral grounds, nonetheless presents today's administrator with myriad problems. Emotionally charged controversies surrounding such concerns as busing; transfer of faculty from school-to-school and, in some cases, from one subject area or one type of position to another; the closing of some schools with close community ties; parental concern for the safety of their children and for "standards" in the schools; and many similar problems—some imagined and some real—have become common, "everyday" problems for educational leaders in all parts of the country.

Other current "spectres" which haunt school administrators include the concepts of "accountability" and performance contracting and the fear of domination by the computer, the systems engineer, and other technological offspring.

Many of the phenomena which are listed here as "problems" are, of course, "blessings in disguise." For example, the current public awareness and concern over education seem, to these writers, far more desirable than the general apathy which has existed for years in many school districts. Student unrest, in many cases, may well be a manifestation of a heightened interest in education and more critical thinking on the part of our youth.

The willingness of teachers to "stand up and be counted" on important issues—even if this sometimes results in teacher actions not popular with, nor condoned by school boards and some administrators—seems a great improvement over a teaching force noted for docility, indecisiveness, and political naiveté.

The crisis in the financial support of public education may well result in a desired redistribution of financial support for schools between local school districts and state and federal governments. School administrators, and others, may become increasingly conscious of the "cost effectiveness" of educational programs with such serendipity as improved long-range planning, program evaluation, and the like.

Aside from the obvious moral, legal and social factors, school desegregation has, either directly or indirectly, spawned increased federal aid to local school districts, and the serious and long overdue reevaluation of curricula, testing programs, teaching strategies, and other shibboleths.

These introductory remarks are not intended, of course, to depict the role of the school administrator as either hopeless or helpless. On the contrary, educational leadership has never been more vital to the success of public education than is true today. And despite existing constraints which seemingly mitigate against effective educational leadership, today's administrator has far more resources at his disposal than at any time in our history. He is better prepared, personally, and has a greater number of professional, highly specialized assistants or resource people to call upon. And as Robert Havighurst[6] has recently pointed out, there are indications of a public readiness—even impatience—for strong, responsible leadership.

It is patently evident, however, that the effective administration of public education in the 1970's and beyond requires administrative styles and strategies consistent with the complexity of the institution itself. A traditional, monolithic approach to administrative leadership seems doomed to failure. If, indeed, such an approach has ever been effective—as well as efficient—that day is past. The literature of recent years is equally replete with the tolling of funeral bells for authoritarian, highly bureaucratic administrative organizations and, coincidentally, with strident pleas for *new*, effective approaches to administration. Cries for better theories, clearer definitions of the administrator's role, participatory management, and the development of a "science" of administration—and the list of "remedies" is practically endless—echo throughout the professional journals and texts.

The purpose of this chapter is to focus on one concept—*The Administrative Team Concept*—which, to these writers, shows great promise as an effective administrative strategy.[7]

The values of group decision making are well documented in the literature of the behavioral sciences.[8] Only in recent years, however, has the administrative team concept begun to appear in the professional literature.

---

6. Robert L. Havighurst, "Educational Leadership for the Seventies," *Phi Delta Kappan*, 53 (March 1972): 403-406.

7. We use the term *strategy* here as Douglas McGregor referred to it in *The Professional Manager* (McGraw-Hill, 1967). McGregor made a distinction between *style* and *strategy*. He refers to managerial *style* as a ". . . method of coping with organization reality which evolves out of trial and error and is not deliberately adopted or even fully recognized by the individual." (p. 68) *Strategy*, on the other hand, is deliberately planned and adopted.

8. Thomas Gordon, *Group-Centered Leadership* (Boston, Massachusetts: Houghton-Mifflin Co., 1955). Rensis Likert, *New Patterns of Management* (New York: McGraw-Hill Book Co., 1961). James G. March and Herbert Simon, *Organizations* (New York: John Wiley & Sons, 1958). Robert L. Saunders, Ray C. Phillips, and Harold T. Johnson, *A Theory of Educational Leadership* (Columbus, Ohio: Charles E. Merrill Books, Inc., 1966). Herbert Simon, *Administrative Behavior*, 2d ed. (New York: The Macmillan Company, 1957).

Even today, little research data are available to document—or disprove—the advantages of this management strategy. Perhaps the dearth of definitive work in this area can be at least partly explained by the following quote taken from a 1971 publication of the prestigious American Association of School Administrators.

This publication concentrates on profiles of cabinet-level administrative positions rather than on a broader definition of the administrative team, *because the management team concept means different things to different people.*[9] (emphasis mine)

In this excellent publication the importance of the administrative team is clearly recognized. The responsibilities, functions, and working relationships of the several assistant-superintendent level administrators found in the typical school district are identified along with recommended modifications in their administrative activities. The "administrative team," as a concept, however, remains undefined.

Reviews of the literature and discussions with knowledgeable school administrators would lead one to accept the AASA statement concerning the conflicting views of the management team concept. For the purpose of this chapter, however, an administrative team is considered to be *a group of administrative specialists who meet formally under the senior member of the team and who, working together as a team, discharge their decision making responsibilities within the organization.*

This working definition, however inadequate, is designed to point out that the *method of operation* of the team is far more vital to the team concept than is the composition of the team itself.

In these next few pages, an attempt will be made to develop a rationale for team administration. Operating principles for the administrative team will be described in some detail. Finally, implications for the preparation of administrators who can work effectively in such teams will be noted.

PRINCIPLES OF TEAM OPERATION

Team operation depends as much upon the behavior and contribution of every member of the organization as it does upon the Chief Administrator.[10] This proposition is increasingly shared by behavioral scientists as well as by professional education managers who have devoted themselves

9. American Association of School Administrators, *Profiles of the Administrative Team* (Washington, D.C.: The Association, 1971).

10. Gordon L. Lippitt, *Organization Renewals* (New York: Appleton-Century-Crofts, 1969), p. 109.

to the study or practice of research-based administration. The administrative leader *does*, as Halpin[11] suggested, initiate structure and provide consideration, two unusually important leadership processes in the life of the organization. Yet without the full commitment and contribution of every organizational member, management is seriously hampered in its efforts to build a truly viable organization. The realization of this critical role of teamwork has prompted educational managers to seek new ways to involve every member of the organization. As Likert commented:

"Management will make full use of the potential capabilities of its human resources only when each person in an organization is a member of one or more effectively functioning work groups that have high degrees of group loyalty, effective skills of interaction and high performance goals."[12]

It is through team operation that renewal, the concept which is rapidly becoming a household word in management circles, can be realized.[13] For the individual, for the organization, for all of the social order, renewal offers perhaps man's last opportunity to achieve viability and stability in an all-too-rapidly changing world. At a time when man-made groups are filled with stress, dissention, conflict and even revolt, renewal just may be the most effective strategy for educational stability and advancement.

Conditions Essential For Team Operation

Effective team operation is not a simple and easy task to accomplish. For many organizations, it never becomes a reality—at least one may conclude it never operates at its highest level of effectiveness. It is the kind of goal for which every group member including the administrative leader must constantly strive. It seems to be a sort of elusive goal which denies any one set of principles which work equally well for every organization. Behavioral scientists have, however, recommended for consideration several concepts and related principles which, when adapted to the particular organization, can be very helpful in establishing conditions which foster team operation. Six such concepts are leadership, climate, communications, needs, decision making, and creativity. Each of these six concepts with related principles of team behavior are given more extensive treatment in the pages which follow.

11. Andrew W. Halpin and B. James Winer, "The Leadership Behavior of The Airplane Commander," mimeographed (Columbus, Ohio: Ohio State Research Foundation, 1952).
12. Rensis Likert, *New Patterns of Management* (New York: McGraw-Hill Book Co. Inc., 1961), p. 99.
13. Lippitt, *op. cit.*, p. 98.

Leadership

Up until the last decade or so, leadership research dealt largely with studies of leadership traits: courage, intelligence, physical prowess, decisiveness, considerateness and other similar characteristics. No consistent pattern of traits emerged from the many studies conducted. One cannot conclude that there is a particular group of persons who are destined for leadership regardless of the organizational circumstances in which they find themselves. Team operation or management thus does not depend heavily upon having one particular kind of administrative leader at its helm.

Similar to research findings about leadership as traits, later studies on the "great man" theory have offered inconclusive results. Stogdill found that, while the typical great leader is more controlled, many successful ones are lacking in emotional control. Effective leaders may and often do, have widely varying personalities.[14]

More recent leadership research has centered upon the situation in which leadership occurs. The leader and other members of the group interact at a particular time and a particular place and within a specific set of circumstances. In this situation true leadership emerges, that behavior which tends to help the group reach its goals. As the situation changes, different leadership abilities are required. One who exerts leadership must be adaptive and flexible so as to be able to respond immediately to each unique set of circumstances. This latter view of leadership as a function of the situation appears to offer more appropriate support for team operation. Teamwork tends to be most effective when there is a sharing of leadership.[15] The administrator who devotes as much effort toward the identification, development and utilization of leadership resources of each member of the group as he does toward exercising his own leadership perogatives will enhance the operation of team management.

The educational administrator who sees this importance of each staff member would tend to organize the resources of the educational institution in such a way that all could work together in defining purposes and objectives. He would utilize the motivation inherent in each person. His work would consist in large measure of removing obstacles, creating opportunities, freeing potential, challenging capacity, and encouraging growth and change. At the conclusion of each learning, growing experience by faculty and staff, he might well have to recall the words of Lao-Tse (600 B.C.):

14. Henry Clay Smith, *Psychology of Industrial Behavior* (New York: McGraw-Hill Book Company, Inc., 1964), p. 246.
15. Lippitt, *Organization Renewals,* p. 109.

"Of the best leaders
the people only know that they exist;
the next best they love and praise;
the next they fear;
and the next they revile.
When they do not command the people's faith,
some will lose faith in them,
and then they resort to recriminations:
But of the best when their task is accomplished,
their work done,
the people all remark,
We have done it ourselves!"

Climate

Team operation depends heavily upon a group of increasingly self-actualizing, fully functioning human beings. An organization built on the assumptions and values of self-actualization is likely to have a climate which is conducive to meaningful group work and thus to team operation.

Many of the significant effects of the organization's climate upon its members and the people it serves have been verified through research. A few of the generalizations which are generally supported in the literature are:

1. Climate is significantly associated with the accuracy of communications in that a hostile, more closed, formal climate tends to block communications while a supportive, relatively open, informal climate with high levels of trust facilitates accuracy in communications.
2. An open and supportive climate produces greater spontaneity and initiative and tends to improve decision making and problem-solving ability of those affected.
3. An open and supportive climate tends to promote greater production and improved quality on both voluntary and required tasks.
4. An open and supportive climate tends to build greater commitment to and enthusiasm for the philosophy and mission of the organization.
5. A closed and hostile climate on the other hand, tends to produce greater dependence upon the formal leaders and less upon group member initiative.
6. A closed and hostile climate tends to promote self-interest, lack of creativity, and lack of enthusiasm for the work of the organization.
7. A closed and hostile climate produces greater anxiety and passiveness toward the organization and its mission.

8. A closed and hostile climate often produces defensiveness and a resistance to ideas and suggestions, whereas a supportive climate tends to produce outgoing, open, tolerant employees and clientele.[16]

Many factors appear to be associated with the establishment and development of the organization's climate. Community support, organizational resources, needs fulfillment, and opportunity for recognition and advancement tend to influence the quality of the organization's climate. One of the most influential factors—one which appears to be most enhancing of the climate conducive to team operation—seems to be the beliefs of the administrative leader of the team.

A number of researchers at the University of Florida investigated the perceptual differences between good and poor professional workers or helpers in teaching, counseling, and the ministry.[17] From these studies, good helpers were clearly distinguished from poor ones with respect to the following beliefs about people.

1. Internal over External Frame of Reference—The true or good helper's general frame of reference can be described as internal rather than external. He is sensitive to and concerned with how things seem to others with whom he interacts and uses this belief as a basis for his own behavior.

2. People over Things Orientation—He is most concerned with people and their reactions rather than with things, events, and activities.

3. Meanings over Facts Orientation—The good helper is keenly sensitive to how things seem to people rather than concerned exclusively with concrete events. The exploration of ideas and discovery of meanings through interactions with people are important components of his belief.

4. Immediate over Historical Causation—He seeks the causes of people's behavior in their current thinking, feeling, beliefs, and understandings rather than in objective descriptions of the forces exerted upon them now or in the past.

5. Able over Unable—The true helper perceives others as having the capacities to deal with their problems as opposed to doubting their capacity to handle themselves and their lives.

16. Dale Alam, "Summary of Seminar Proceedings in Education 700," mimeographed (Gainesville, Florida: University of Florida, 1964).

17. Arthur W. Combs, *The Professional Education of Teachers* (Boston, Massachusetts: Allyn and Bacon, Inc., 1965).

6. Friendly over Unfriendly—He sees others as being friendly and enhancing, not a threat but well-intentioned rather than evil-intentioned —"on our side."
7. Worthy over Unworthy—He tends to see other people as being worthy of our respect. They are seen as possessing dignity and integrity which must be respected, rather than seeing them as unimportant.
8. Internal over External Motivation—The helper tends to see people and behavior as developing from within rather than as products of external forces. People are seen as dynamic, creative, and growing rather than passive, inert, and vegetating.
9. Dependable over Undependable—He tends to see people as basically trustworthy and dependable in the sense of behaving in a lawful way.
10. Helpful over Hindering—He tends to see people as being potentially fulfilling and enhancing to self rather than impeding or threatening. He tends to regard people as important sources of satisfaction rather than sources of frustration and suspicion.

Communications

Team operation requires trust and openness in communications and interpersonal relationships. A free and trusting relationship will encourage open and frank communications.[18] A high level of tolerance for different opinions and personalities will be manifested. Emphasizing the trust element in open communication, Argyris[19] suggested the importance of "authentic" communications. Nothing appears to hamper team operation more than insincerity in human relationships.

Hicks supports the significance of open communications to society by commenting:

Communication not only raises man from the level of a three-year old toward becoming a full person, it also provides the foundation of human interaction. Any time one human interacts with another, some form of communication takes place.[20]

Communications may be defined as a process by which ideas are originated, passed along, received and related to in such a way as to result in formation of opinions and attitudes. Communication is thus viewed as

18. Lippitt, *Organization Renewals*, p. 113.
19. Chris Argyris, "The Individual and Organization: Some Problems of Mutual Adjustment," *Administrative Science Quarterly*, Vol. 2, No. 1 (1957), p. 10.
20. Herbert Hicks, *The Management of Organizations* (New York: McGraw-Hill Book Company, 1967), p. 281.

a continuous, free-flowing process. In order to improve team operation through open communications, one needs to develop some understanding of the process. One acceptable model of communications congruent with the above definition is that by David Berla.[21] Elements included in Berla's model are:

1. Communication source—originates from an individual or a group with a purpose which needs to be communicated.

2. Encoder—part of communication process by which the ideas of the source is put into some code or special language. Usually involving the individuals' motor skills such as muscular reaction and vocal transmission, the encoder may be the same person or source or sender or it may be a "go between." Often this second hand encoding may account for communications breakdown.

3. Message—translation of the idea into some symbolic code which usually takes form in language.

4. Channel—linkage between the source of the message and the target of the communication. It includes all persons and processes between these two elements.

5. Receiver—target (person or persons) on the receiving end of the communication.

6. Decoder—part of communication process by which the message is put into some form by which it can be used by the receiver. Decoding as was the case with encoding may be accomplished by the receiver or another person or group assisting with the process.

7. Meaning—interpretation given the communication by the receiver. It may or may not coincide with the intended message by the communication.

8. Noise—element present in all aspects of communications and which tends to interfere with the message, reducing its accuracy.

9. Feedback—response which the receiver makes upon receiving and interpreting the meaning of the sender's encoded message.

As evident by Berla's analysis, breakdowns and barriers may occur at many points in the communications process. Among the most prevalent causes of communication breakdown are language problems, excess noise, faulty channels, inaccurate translation and interpretation, and premature evaluation. The paramount importance of inaccurate translation and interpretation is emphasized by Koontz and O'Donnell.

21. David K. Berla, *The Process of Communication: An Introduction To Theory and Practice* (New York: Holt, Rinehart and Winston, 1960), p. 23.

It is often not enough to pass on a communication word for word; either it must be put into words appropriate to the framework in which the receiver operates or it must be accompanied by an interpretation which will be understood by the receiver.[22]

There are several implications from the model of communications for the organization which would improve team operation. Communications will be more accurate and, consequently, team operation enhanced if each member:

1. Understands the communication process, the points at which most breakdowns occur and the factors which most often impede accuracy.
2. Understands the critical contribution of open communications to morale, decision making, interaction, cooperation and goal-directed behavior.
3. Understands the nature and importance of prompt and accurate feedback and the prompt and sincere response to feedback. This understanding appears to be especially important for the administrative leader.

Needs

Human beings, with their myriad of complex and varying needs, form and join organizations to meet these needs. To deny the presence, importance and fulfillment of human needs by members of the organization is to deny opportunity to build a true team approach in the organization. On the other hand, to fulfill human needs, especially those at lower levels, is to free members of the organization to devote their full energy to the purposes and goals of the organization.

A pioneering effort in the study and understanding of human needs was made by Maslow[23] in his hierarchy of needs. As lower levels of needs are met, higher level needs become increasingly important to the person. Finally at the apex are self-actualization needs, the fulfillment of which appear to enhance the effectiveness of team operation. Maslow characterized self-actualizing or fully-functioning persons as:

1. Efficiently perceiving and being comfortable with reality
2. Accepting of self, others and nature
3. Spantaneous
4. Problem-centered
5. Detached (Disinterested or non-personal)

22. Harold Koontz and Cyril O'Donnell, *Principles of Management*, 3d ed. (New York: McGraw-Hill Book Company, 1964), p. 599.
23. Abraham H. Maslow, "Self-Actualizing People—A Study of Psychological Health," Personality Symposium No. 1, 1950.

6. Autonomous and independent from culture and environment
7. Continually and freshly appreciative
8. Effective in interpersonal relations
9. Empathetic for mankind

The characteristics of the fully-functioning person appear to associate positively with good team operation. Consequently, when the organization provides for the meeting of this highest level of needs of each of its members, creative energy is released to accomplish the goals of the team. A creative team can release its full physical and mental resources. With the motivation and courage to let themselves go, try the new and innovative, tremendous progress can be made.[24]

McGregor[25] offers the manager an opportunity to fuse individual needs and goals with organizational needs and goals with his theory "x" and theory "y." The concept advanced in theory "x" is that man is lazy, passive, unwilling to assume responsibility and resistant to change. He therefore must be controlled and directed by the manager. The concept advanced in theory "y" is that man does wish to grow, is active and wishes to maximize his usefulness in the organization. The manager can therefore afford to provide opportunity for greater involvement for every person in planning, setting program goals and working toward the achievement of goals. A summary of the assumptions in theory "y" follows.

1. The expenditure of physical and mental effort in work is as natural as play or rest. The average human being does not inherently dislike work. Depending upon controllable conditions, work may be a source of satisfaction (will be voluntarily performed) or a source of punishment (will be avoided when possible.)

2. External control and the threat of punishment are not the only means for bringing about effort toward organizational objectives. Man will exercise self-direction and self-control in the pursuit of objectives to which he is committed.

3. Commitment to objectives is a function of rewards associated with their achievement. The most significant of such rewards, *e.g.*, the satisfaction of ego and self-actualization needs, can be direct products of effort directed toward organizational objectives.

4. The average human being learns, under proper conditions, not only to accept but to seek responsibility. Avoidance of responsibility, lack of ambition, and emphasis on security are consequences of experience, not inherent human characteristics.

24. Irving R. Weschler, "The Leader Looks at Creativity," *Looking Into Leadership Monographs* (Washington, D.C.: Leadership Resources, Inc., 1961), p. 10.
25. Douglas McGregor, *The Human Side of Enterprise* (New York: McGraw-Hill Book Company, Inc., 1960), pp. 45-48.

5. The capacity to exercise a relatively high degree of imagination, ingenuity, and creativity in the solution of organizational problems is widely, not narrowly, distributed in the population.
6. Under conditions of modern industrial life, the intellectual potentialities of the average human being are only partially utilized.[26]

Decision Making

Decision making is not only the core administrative process but also the most often performed process by all members of the organization. Every organization, to be effective, must develop the ability to make decisions. The decision making process is the single most important criterion by which the effectiveness of the organization is judged, its structure identified, its goals accomplished and its life renewed.

One of the most prominent theories related to decision making is advanced by Griffiths.[27] In his model, decision making is viewed as the central process in administration. The chief purpose of administration is that of monitoring decisions made in the organization. Four propositions are presented:

1. The structure of an organization is determined by the nature of the decision making process.
2. If the formal and informal organization approach congruency, then the total organization will approach maximum achievement.
3. If the administrator confines his behavior to making decisions on the decision making process rather than making terminal decisions for the organization, his behavior will be more acceptable to his subordinates.
4. If the administrator perceives himself as controller of the decision making process, rather than as maker of the organization's decisions, decisions will be more effective.

From these propositions it seems clear that it is important for the administrative leader to accept the idea of worker participation in the making of decisions in the organization. It also appears clear that effective team operation is dependent upon the utilization of appropriate steps and guidelines for decision making in the solution of problems.[28] Such appropriate steps and guides were offered by Lippitt in his work with "Y" youth groups.

1. *A clear definition of the problem.* As the problem is unambiguous and relatively easy to understand, the decision making process will be enhanced.

---

26. *Ibid.*
27. Daniel E. Griffiths, *Administrative Theory* (New York: Appleton-Century-Crofts, 1959), pp. 71-91.
28. Lippitt, *Organization Renewals,* p. 112.

Problems which are too general or poorly defined defy the group's efforts to come to grips with them. Problems which are clearly outside the responsibility of the group are obviously discouraged.

2. *A clear understanding as to who has the responsibility for the decision.* Clarification of the group's limits of freedom in dealing with the problem and its degree of responsibility tend to enhance decision making and thus team operation.

3. *Effective communication for idea production.* In the same manner that premature interpretation of a message can block accuracy in communication, first solutions in the solving of problems can result in poor decision making. The group requires time to get ideas out in the open. Brainstorming, small discussion groups and idea identification processes can often be valuable to effective decision making.

4. *Appropriate size of group for decision making.* Large groups can often get bogged down during decision making activity. One solution may be the use of such groups in considering decision alternatives.

5. *A means for effective testing of different alternatives relative to the problem.* The identification, collection and utilization of appropriate data are extremely important to decision groups. Alternatives cannot be realistic with inadequate data and it is unfair to the group to proceed without sufficient data for realistic decision activity.

6. *A need for building commitment into the decision.* Groups function more effectively and efficiently if they are convinced the decision they make has a reasonable chance of implementation. They can build into the decision responsibility and delegation for implementing the decision. Failure to pin down responsibility can result in frustration, apathy and discouragement by group members.

7. *Honest commitment of the manager or leader to the group decision making process.* The administrative leader can make a personal contribution but if it is made too early in the decision process it will at least retard if not prevent the creative idea input by members of the group. His chief concern is the process which is occurring, the growth of members of the group and their increasing level of self-actualization.

8. *A need for agreement on procedures and methods for decision making prior to deliberation of the issues.* Early agreement on decision making procedures can assure greater success with decision outcomes. Lack of such agreement may often lead to "splits" in the group, delays in beginning decision making deliberation and unnecessary confusion in early decision making activity.[29]

29. Gordon L. Lippitt, "Improving Decision-Making with Groups," *Y Work With Youth* (A publication of the Program Services Department, National Council of YMCA's, April 1958).

Creativity

Creativity is often associated with the production of a product such as painting, invention, literary masterpiece or music composition. It is a process resulting in something tangible, something one can feel, see, touch, or hold. Man continues the search for and encouragement of the tangible, but the concept of creativity has been enlarged to include ideas, activities, decisions, relationships, critical thinking, problem solving—results of man's cognitive and affective powers.[30] When viewed in this way, creativity can become the force to turn an otherwise dull and routine school operation into an exciting, challenging, and meaningful experience in learning. The administrator's job can become one of coordination and releasing energies and human resources rather than one of directing and controlling the creative energies of the organization.

Truly effective team operation depends upon the acceptance of this latter view of creativity, a view which encourages the maximum development and utilization of different resources of all individuals in the organization. Often the more difficult problems will require the specialized knowledge and experience of all the people in a given group.[31] Under such circumstances the leader of the team realizes great benefit in knowing the interests and abilities of each member of the team. All too often the person who could contribute best to a particular problem is not asked to help. Many organizations suffer as much from such unused talent as from the misuse of talent. Team operation will require that all of the resources of all of its members be identified and utilized in order to reach the goals of the organization.

A summary of the foregoing principles of team operation is best organized around the characteristics of effective team operation. Among these characteristics are the following.

1. Team operation tends to be enhanced when leadership from the group is identified, developed and shared.

2. Progress toward organizational goals tends to be increased when administrative leadership assumes a major role of removing obstacles, creating opportunities, freeing potential, challenging capacity and encouraging growth and change.

3. An open and supportive climate tends to enhance the development of fully-functioning human beings who, in turn, enhance team operation and group development.

---

30. Arthur W. Combs, ed., *Perceiving, Behaving, and Becoming* (Washington, D.C.: ASCD, 1962).

31. Lippitt, *Organization Renewals*, p. 109.

4. Belief structure of the administrative leader appears to be associated with team operation in that a positive, internal view of life tends to enhance team operation whereas a negative, external one tends to retard team operation.
5. Trust and openness in communications along with a high level of tolerance for different opinions and personalities tend to promote effective team operation.
6. Team operation is improved and communications accuracy increased if the team members:
   a. understand communications processes;
   b. understand and accept the contribution of openness to morale, decision making, interaction, cooperation and goal behavior;
   c. understand and develop commitment to a process of immediate and sincere feedback.
7. Meeting the needs (physical, social and psychological) of members of the group will free creative energy for application toward the organization's goals.
8. Worker participation in the making of decisions in the organization tends to shorten the time required for decision adoption and subsequent team operation or group activity.
9. Utilization of appropriate steps, procedures and criteria in group decision activity will enhance team operation and thus organization goal achievement.
10. True team operation will utilize fully the variety of talents and resources available throughout the organization.

As complex and as complicated as team operation may be, it may be the best solution to the difficult educational leadership problems in today's schools. More than ever, society needs creative organizational leadership, a new breed which searches for new approaches, new strategies, and new visions of that which "might be" in education. Team operation offers that opportunity for new approaches, new strategies and new vision, but more importantly, team operation offers a worthwhile challenge to creative organizational leadership to make it all possible.

ADMINISTRATIVE TEAM—IMPLICATIONS FOR THE SCHOOL SUPERINTENDENT

The administrative team and the strategies implicit in its organization can be discussed in the context of nearly all management systems. This section will deal with the characteristics of an administrative team alignment as it relates to administration and management in the public schools

ADMINISTRATIVE TEAM CONCEPT 93

—an alignment which seeks to provide a more viable and productive management style.

As was pointed out earlier, the last two decades have produced changes in the operation and direction of public schools unparalleled in the brief history of American education. School superintendents have adopted and adapted classical management styles in an attempt to organize schools in a systematic and efficient pattern. Most often these patterns of organization have followed a "two-stream" model with a major distinction or separation between instructional and business functions. Further sub-divisions of these two major areas vary from school district to school district depending upon size and other factors; most however, follow a format similar to that of the following diagram.

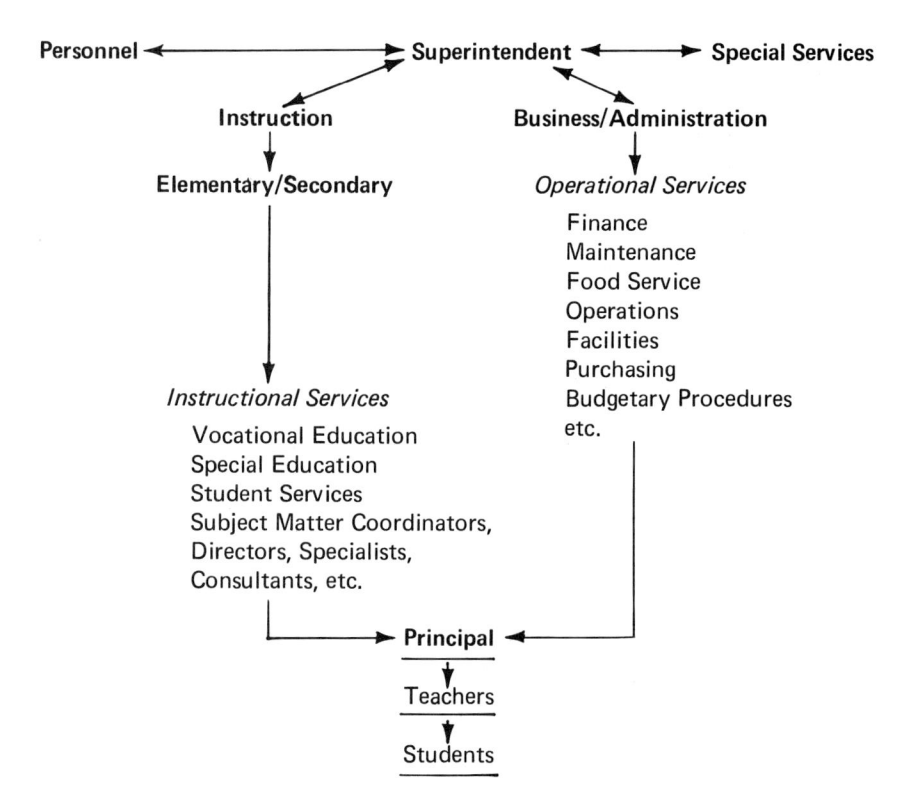

```
Personnel ←――――――→ Superintendent ←――――→ Special Services
                    ↙           ↕
            Instruction      Business/Administration
                ↓                   ↓
       Elementary/Secondary    Operational Services
                │                  Finance
                │                  Maintenance
                │                  Food Service
                │                  Operations
                │                  Facilities
                ↓                  Purchasing
       Instructional Services     Budgetary Procedures
         Vocational Education     etc.
         Special Education
         Student Services
         Subject Matter Coordinators,
         Directors, Specialists,
         Consultants, etc.
                 └――――→ Principal ←――――┘
                            ↓
                         Teachers
                            ↓
                         Students
```

As suggested by the diagram, all of the management functions typically converge at two levels; the superintendency and the principalship. The superintendent is responsible for decisions which effect the district although he himself does not typically make all of them. The principal, teachers and students are subject to those decisions. In recent years, efforts have been made to relocate decision making authority closer to those influenced by the decisions. The problems associated with relocating this authority and responsibility for decision making are worthy of consideration but not the focus of this discussion. Problems centering around the convergence of decision making authority at the level of the superintendency are, however, presented and discussed in some detail.

As noted in the diagram above, decisions relating to instruction and administration as well as other special services are either made by or reported directly to the chief administrator in most districts.

In addition, a number of service agencies usually have direct access to the superintendent's office providing information on the one hand and requiring guidance on the other. Two of the more common service divisions are personnel and data processing with increasing emphasis now placed on the development of research and development as a special service arm in many school districts.

Conventional line/staff administrative organizations have proved less than effective in meeting the emerging problems facing the schools. The need for creative new management strategies to meet challenging new problems is evident. It should be noted however that perhaps the most common reason for failure of any administrative structure is failure of the chief administrator to follow good management practices.

1. People with too little talent are promoted beyond their ability to perform. (Peter Principle)

2. People with much talent are not permitted to assume responsibility commensurate with their ability. (In some cases these people leave the organization or simply settle back to function at a level below their capacity.)

3. Too much or too little attention is paid to "rule of thumb" estimates regarding span of control. Obviously, span of control guidelines are without basis unless personal factors are taken into consideration. Undoubtedly many schools do not review their organizational charts, personnel assignments and functional assignments often enough to determine whether assignments are realistic and effective. Too often new divisions are organized, positions created and personnel added without benefit of a penetrating analysis and evaluation of the existing administrative structure.

4. Chief administrators depend too little and sometimes (although rarely) too much on their administrative staff for recommendations.

5. Chief administrators delegate responsibility but not authority to carry out assignments.
6. Assignments are not clearly defined thus resulting in unnecessary competition between divisions/departments which may dull the effectiveness of the organization.

Obviously, there are many other equally significant reasons why administrative organizations have failed in their efforts to promote effective and efficient management.

School superintendents have, in the past, recognized several of these problems and reassigned personnel, reorganized departments and restructured organizational charts in an attempt to make schools more responsive to their clientele. Unfortunately, however, many of these changes have not provided long-lasting vitality to the organization. Indeed, a key consideration of, and prerequisite to, effective administration is the appointment of the most capable people for each position. Consideration of other basic management principles including span of control, consistency and specificity of assignment, unity of command, commensurate authority and responsibility, appropriate delegation of both authority and responsibility, balance and organizational structure are also important considerations.

While these factors are extremely important in the development of a management strategy and structure, one must not become too enamored with the effectiveness of these and other corporation management principles. While many verbalize the virtues of efficiency in business, one must also be cognizant of the fact that schools are different in several important ways. Schools are not, for example, private organizations; they do not generate a profit and they are typically under the control of a board of directors who know little about operations, technology or product of the enterprise. In addition, schools are generally subject to restrictive budgetary controls and unable to consider the possibility of increasing product cost in exchange for improvements in product quality.

There is less parallel in management of public schools and private corporations than many would have us believe. As a matter of fact, it should be counted as some sort of miracle that public schools have maintained themselves, using borrowed management styles, so well for so long under such adverse conditions.

Obviously, the intelligent school administrator cannot expect to operate schools successfully for any extended period of time using either his own judgment exclusively *or* unmodified corporate management practices. He must recognize and adjust to several factors including a recognition that schools:

a. Provide service to only a few of those expected to bear the cost of the service;
b. Require centralized and consolidated administration and organization in an enterprise which relates with its publics on a very personal basis;
c. Place major responsibility for interpreting the school to its publics on personnel who are placed annually in an adversary relationship with management through the collective negotiations process; and
d. Have few of the positive characteristics of private business, most of the negative attributes of a public utility and are expected to operate as though they combine the best qualities of each.

The effective school superintendent of the 70's must not only surround himself with high quality professional personnel—he must find effective ways of encouraging, soliciting and utilizing their counsel.

The formation of administrative cabinets has been a demonstrable positive step in this direction. Too often, however, these arrangements have resulted simply in an inner organizational administrative structure which functions as a reporting agency rather than a decision making body. Crucial decisions relating to instructions are too often made in offices not completely familiar with the instructional programs of the district. The fact that acceptance of a low bid sometimes results in subtle (and sometimes not too subtle) shifts in program emphasis has often been recognized—less often corrected. A recognition that efficiency is dependent upon a harmonious interfacing of resources committed to staffing, facilities, technology and curriculum and that these decisions require a much more complex and intricate communications system if the quality of school programs is to be optimized.

How Does the Team Operate?

The administrative team concept as related to the school superintendency is indeed more a state of mind than a definable organizational structure. It is somewhat simpler to describe the relationships between members of the team than to place descriptions or titles on the positions of its constituents.

The most obvious description of an administrative team is that it functions as a team. More specifically, the members of the team *cannot* operate independently. This is true for two reasons; in a truly functional team, individual team members cannot take action independently since they know such action could influence operations and force decisions (often prematurely), in other areas. Secondly, it is considered likely that no one member of the team has sufficient information to make decisions which could affect the total operations of the school system.

This is not to say that administrators functioning as a part of the administrative team divest themselves of either the authority or responsibility for making decisions. On the contrary, decisiveness and organizational ability remain very much the descriptors of key administrative personnel. Normal operational decisions which are unique to the various subdivisions of the school district should and must be made as needed.

The conditions for team decision making outlined above relate to major policy issues which face school districts on a regular basis, not day-to-day concerns. Those key questions which confront school districts in the areas of planning, finance, personnel policies, accountability, instructional program design and public relations are examples of concerns which require the composite consideration of the administrative team. Obviously it is impossible to describe all of the specific problems which might require such attention in any district. Indeed, one of the major tasks of a team is to identify, in advance, areas of concern which merit the study and attention of the group.

By taking advantage of the positive elements present in the organization of an administrative team, it is clearly predictable that decisions which are reached will be viable and more likely, acceptable. In addition, decisions and recommendations can be implemented quickly and with a higher probability of success.

It has been argued with convincing evidence that effective communication is the key to successful school administration. The administrative team organization provides the forum for such communication. As outlined earlier in this chapter, team members have a responsibility to assure that all sides of each issue brought before the team are thoroughly discussed. Dissident attitudes and positions must be encouraged and all information considered impartially on the basis that the ultimate decision is not a foregone conclusion.

Several factors appear to be critical to the successful functioning of an administrative team. They include:

a. Scheduling of regular meetings—Meetings must be held on a regular basis. Even if the school district does not have significant problems on the horizon, the team should meet to exchange information, "brainstorm," set future goals and evaluate previous decision making.

b. Staff work should be detailed—One of the important factors in good decision making is having available as much information as possible prior to making a decision. Members of the team should participate actively in the development of position papers pro and con, relative to any question under consideration. Team members should alternate in playing "devils advocate" to positions which appear to be clear-cut.

98 ADMINISTRATIVE TEAM CONCEPT

c. Written agenda—Special efforts must be made to develop written agenda materials for meetings. Agenda building will require considerable time; however, one of the most important functions of the team is to plan sufficiently in advance to head off major concerns before they become problems. Tentative agenda items should be identified three to six months in advance. This will provide sufficient time for necessary staff work prior to consideration of an item. The agenda should, however, remain flexible enough to deal with special concerns which may arise on a week to week basis. Rigid adherence to the agenda when problems emerge unexpectedly is to disregard the major role of the team—participation in decision making.

d. Records of team meetings—Adequate records should be made of each team meeting with special attention focused on matters of accord, disposition of items, and special assignments to members of the team. Of particular importance are those items which will require broader dissemination to employees of the district in the form of procedural memoranda or items which require change in school policy.

e. Team decisions—Needless to say, one of the most critical determiners of the team's success is the extent to which the total membership of the team will commit itself to a position taken by the team. Unless all of the members of the team are willing to support decisions reached cooperatively by the team, there can be no effective functioning of the team. Members of the team who bring a position of advocacy to the deliberations must be willing to submerge that position once deliberations have been completed and support with equal vigor the position developed and agreed upon.

Who Belongs to the Team?

The decision as to which personnel will constitute the membership of the team is made by the superintendent. This decision must be made based upon several factors including: size of school district, internal organization of the district, personnel considerations and understanding of group process.

In a smaller district the superintendent may include his entire administrative staff which may be limited to one additional central office administrator and two or three building level administrators.

In medium or large size districts, the superintendent may wish to include only personnel at the level of deputy and associate superintendent. In any event, those persons who have major responsibility for decision making which affects large numbers of people or significant functions within the district should be represented. Staff members who have responsibility for finance, personnel and instruction must, obviously, be included in team deliberations. The extent to which other service divisions of the district are included or represented will depend, to a large extent, on the considerations noted above with respect to internal organization and principles of group dynamics.

Effective Team Leadership

This discussion would not be complete without some attention to the role of the superintendent in the administrative team. Actually little needs to be said in this regard. Simply stated, the concept of the administrative team cannot succeed without his full commitment. The superintendent has the legal and ethical responsibility to provide leadership to a school district. He alone can make the decision to delegate authority; he cannot escape the responsibility of the position.

Only he is in a position to maintain and nurture the climate of shared decision making. He must maintain an atmosphere which encourages the presentation of alternative viewpoints in all areas of discussion. This responsibility need not extend to a concensus management mode, but must project a guarantee that all items of discussion will be considered impartially and discussed completely. Beyond that, a superintendent must be willing to modify his position if logical and compelling evidence supports a counter position. Without these characteristics, a superintendent cannot and should not expect the administrative team to reach its maximum effectiveness. Certainly there are many other reasons why an organization of the administrative staff as proposed here might fill. Obviously, it is possible that a majority of the administrative team as well as the superintendent might consistently exercise bad judgment. Cooperative decision making is not an easy task under the best of circumstances. Most administrators have had little, if any, experience in shared decision making. Superintendents may be reluctant to delegate their authority. Team members may not believe that the superintendent is sincere in his efforts to achieve a shared decision making model. The reasons for failure are manifest.

One might point out again, however, that education is faced with an increasing number of problems which extend the creative abilities of all administrators and exceed the capabilities of most individuals.

It would appear that the potential for more effective and creative decision making lies in patterns of management and administration which can marshall the talents of several rather than a single administrator. This is the hopeful promise of the administrative team concept.

THE PREPARATION OF ADMINISTRATORS

The administrator who can work effectively in the team setting does not just "happen." While the "leaders are born, not made" theory has long since passed from the scene in administrative preparation circles, many—if not most—preparation programs still lag behind what is known about the administrative process.

It is probably true that preparation programs have changed over the years, just as the concept of the role of the administrator has changed. Early in the century, reflecting society's concern with productivity and technology, *efficiency* became a central concern of administration and administrative preparation program. Taylor's classic work, *The Principles of Scientific Management*,[32] prompted strong pressures for increased efficiencies in education. The reader interested in a detailed analysis of the scientific management movement and its effect on educational administration can find a comprehensive treatment in Raymond Callahan's *Education and the Cult of Efficiency*.[33]

Following the *efficiency* period in education administration, the emphasis swung to what is referred to as the *human relations* period. This view—certainly supported, if not prompted, by the now-famous Hawthorne experiments, represented a movement toward cooperation and a "democratic," people-centered administration. Here again, this movement was spawned by changing social values in society.

Herbert Simon's writings of the late 40's[34] preceded what can be called the *scientific* era of administration. Simon's work, supported by that of Barnard[35] a decade earlier, called attention to the lack of a core of sound, proven principles of administration. New theories of administrations, based on scientific principles began to appear in the literature.[36, 37, 38, 39]

32. Frederick Taylor, *The Principles of Scientific Management* (New York: Harper & Brothers, 1911).
33. Raymond E. Callahan, *Education and the Cult of Efficiency* (Chicago: University of Chicago Press, 1962).
34. Herbert Simon, *Administrative Behavior* (New York: The Macmillan Company, 1945), and *Administrative Behavior: A Study of Decision-Making Processes in Administrative Organization* (New York: The Macmillan Company, 1947).
35. Chester I. Barnard, *The Functions of the Executive*, 30th anniv. ed. (Cambridge, Mass.: Harvard University Press, 1968).
36. Roald F. Campbell and Russell T. Bregg, eds., *Administrative Behavior in Education* (Chicago: Midwest Administration Center, University of Chicago, 1957).
37. A. P. Coladarci and Jacob W. Getzels, *The Use of Theory in Educational Administration* (Stanford California: Stanford University Press, 1955).
38. Griffiths, *Administrative Theory*.
39. Andrew W. Halpin, ed., *Administrative Theory in Education* (Chicago: Midwest Administrative Center, University of Chicago, 1958).

The attempts to develop a universally accepted theory of administrative behavior are still in process. But while *the* theory of educational administration has not yet evolved, the movement toward theory development continues and has contributed greatly to the existing body of knowledge in educational administration theory.

The student of educational administration today is likely to be enrolled in a program which, if not theory-based, at least includes formal course work in administrative theories. A portion of his studies will probably be selected from the social and behavioral sciences. Courses in public administration, sociology, business, political science, social psychology, and industrial engineering are commonplace in even minimum, masters degree level programs.

Increasing numbers of his classmates, especially in post-masters programs, are likely to have come from non-educational backgrounds. Just as the study of the behavioral sciences has become an accepted and vital segment of many preparation programs,[40] admissions barriers to those without school experience have been lowered. The experience of the engineer, the businessman, the manager, the political scientist, and others are being recognized as appropriate experiences upon which to base the study of educational administration.

The student of educational administration is almost certain to become involved in an increasing amount of laboratory and clinical studies. Extensive utilization of case studies, simulations, T-groups, internships, externships and other clinically related experiences are commonplace. The traditional preparation program composed of a series of formal courses culminating in an examination and a "library-bound" research dissertation is rapidly fading into obscurity.

The preparation program for the administrator who is to work effectively in a team administration environment can be built around the program just described. In order to combat over-specialization and the communication gaps which inevitably ensue, the team administrator should be prepared in breadth as well as depth. A strong cognate area of study in the behavioral sciences seems especially appropriate.

The development of communications skills is, of course, a vital prerequisite for such an administrator, and the team administrator must realize that listening is as much a communication skill as is telling.

Another vital ingredient is a sound base in a theory of decision making which recognizes that decision making is a rational, cooperative

---

40. Farquhar, in a 1970 ERIC publication entitled *The Humanities in Preparing Educational Administrators,* predicts that a slow but steady trend toward an increased use of the humanities as well as the behavioral sciences in administrative preparation programs.

process and that *good* decisions are more likely to be made in a climate which encourages wise group interaction.

The extensive use of laboratory and clinical experiences can serve to strengthen the preparation of the future administrative team member. These activities lend themselves quite readily to the grouping or teaming of participants. By proper utilization of simulations and other laboratory experience, the prospective administrator can become adept at team problem-solving and decision making as a planned outgrowth of his preparation program.

In summary, preparation programs for school administrators must prepare managers for tomorrow's schools, not yesterday's. If, indeed, the concept of team administration is a viable one and a legitimate substitute for the monocratic model, then preparation programs should reflect this management strategy. Current trends in graduate education indicate a gradual move in this direction.

SUMMARY

Complexity and rapid change have become as expected—and accepted—as the proverbial "death and taxes." There is no reason to suspect that society and, concomitantly, its institutions, will not continue to change at, at least, the present rate. The institutions through which society expresses its values and maintains itself will, we believe, become increasingly complex. There appears to be adequate and compelling evidence to support this thesis.

Just as institutions change and become more complex, so must the processes by which they are administered—and so must the men and women who manage these decision making processes. In this brief chapter, these writers propose one alternative to the traditional, monolithic style of educational management. Team administration, while certainly no panacea for all the ills which beset education, appears to offer a viable management alternative which provides for the needs of both the organization and the actors within the organization. No claims are made which would support the team concept as a "theory" of administration. On the contrary, team administration is a frame of reference or an operational strategy which, to these writers, fits neatly within the broader concept of participatory management.

While a working definition of team administration was proposed and several principles of team operation suggested, neither the definition nor the list of principles present claims of completeness.

The experience of these writers suggests that hundreds of school systems throughout the country have created administrative organizations which provide for management teams, administrative councils, executive committees, superintendent's cabinet, and the like. However, the "calling together" of a number of administrative officials for periodic meetings no more insures the administrative team operation proposed in this chapter than would the bringing together of a collection of individuals necessarily result in a "group."

If the principles outlined in this brief chapter are followed, however, these writers are convinced that the decision making process within the school system can be improved. As George Ordiorne recently wrote, "Wrong decisions make more mischief than a thousand devils working their fiendish schemes."[41] While team administration provides no guarantee against "wrong" decisions, there is much to recommend its adoption as the educational management strategy of the 70's.

Bibliography

Alam, Dale. "Summary of Seminar Proceedings in Education 700." Mimeographed. Gainesville, Florida: University of Florida, 1964.

American Association of School Administrators. *Profiles of the Administrative Team*. Washington, D.C.: The Association, 1971.

Argyris, Chris. "The Individual and Organization: Some Problems of Mutual Adjustment." *Administrative Science Quarterly* Vol. 2, No. 1 (1957): 10.

Barnard, Chester I. *The Functions of the Executive*. 30th anniv. ed. Cambridge, Massachusetts: Harvard University Press, 1968.

Berla, David K. *The Process of Communications: An Introduction to Theory and Practice*. New York: Holt, Rinehart and Winston, 1960.

Callahan, Raymond E. *Education and the Cult of Efficiency*. Chicago: University of Chicago Press, 1962.

Campbell, Roald F., and Gregg, Russell T., eds. *Administrative Behavior in Education*. Chicago: Midwest Administration Center, University of Chicago, 1957.

Coladarci, A. P., and Getzels, Jacob W. *The Use of Theory in Educational Administration*. Stanford, California: Stanford University Press, 1955.

Combs, Arthur W., ed. *Perceiving, Behaving and Becoming*. Washington, D.C.: ASCD, 1962.

Combs, Arthur W. *The Professional Education of Teachers*. Boston, Massachusetts: Allyn and Bacon, Inc., 1965.

Fantini, Mario D., and Young, Milton A. *Designing Education for Tomorrow's Cities*. New York: Holt, Rinehart and Winston, 1970.

Farquhar, Robin H. *The Humanities in Preparing Educational Administrators*. Eugene, Oregon: The ERIC Clearinghouse on Educational Administration, University of Oregon, 1970.

Glasser, William. *Schools Without Failure*. New York: Harper and Row, 1969.

---

41. George S. Ordiorne, *Management Decisions by Objectives* (Englewood Cliffs, New Jersey: Prentice-Hall, Inc., 1969), p. 1.

Gordon, Thomas. *Group-Centered Leadership.* Boston, Massachusetts: Houghton Mifflin Company, 1955.

Griffiths, Daniel E. *Administrative Theory.* New York: Appleton-Century-Crofts, Inc., 1959.

Gross, Ronald, and Gross, Beatrice, eds. *Radical School Reform.* New York: Simon and Schuster, 1969.

Halpin, Andrew W., ed. *Administrative Theory in Education.* Chicago: Midwest Administrative Center, University of Chicago, 1958.

Halpin, Andrew W., and Winer, B. James. "The Leadership Behavior of The Airplane Commander." Mimeographed. Columbus, Ohio: Ohio State Research Foundation, 1952.

Havighurst, Robert J. "Educational Leadership for the Seventies." *Phi Delta Kappan* 53 (March 1972): 403-406.

Hicks, Herbert. *The Management of Organization.* New York: McGraw-Hill Book Company, 1967.

Koontz, Harold, and O'Donnell, Cyril. *Principles of Management.* 3d ed. New York: McGraw-Hill Book Company, 1964.

Likert, Rensis. *New Patterns of Management.* New York: McGraw-Hill Book Company, 1961.

Lippitt, Gordon L. "Improving Decision-Making with Groups." *Y Work With Youth* (A publication of the Program Services Department, National Council of YMCA's), April 1958.

Lippitt, Gordon L. *Organization Renewal.* New York: Appleton-Century-Crofts, 1969.

March, James G., and Simon, Herbert. *Organizations.* New York: John Wiley and Sons, 1958.

Maslow, Abraham H. "Self-Actualizing People—A Study of Psychological Health." Personality Symposium No. 1, 1950.

McGregor, Douglas. *The Human Side of Enterprise.* New York: McGraw-Hill Book Company, Inc., 1960.

McGregor, Douglas. *The Professional Manager.* New York: McGraw-Hill Book Company, Inc., 1967.

Ohm, Robert E. "The School Administrator in 1985." in *Educational Futurism 1985*, chapter 6, pp. 93-108. Berkeley, California: McCutchan Publishing Corporation, 1971.

Ordiorne, George S. *Management Decisions by Objectives.* Englewood Cliffs, New Jersey: Prentice-Hall, Inc., 1969.

Saunders, Robert L.; Phillips, Ray C.; and Johnson, Harold T. *A Theory of Educational Leadership.* Columbus, Ohio: Charles E. Merrill Books, Inc., 1966.

Silberman, Charles E. *Crisis in the Classroom: The Remaking of American Education.* New York: Random House, 1970.

Simon, Herbert. *Administrative Behavior.* 2d ed. New York: The Macmillan Company, 1957.

Simon, Herbert. *Administrative Behavior: A Study of Decision-Making Processes in Administrative Organization.* New York: The Macmillan Company, 1947.

Smith, Henry Clay. *Psychology of Industrial Behavior.* McGraw-Hill Book Company, Inc., 1964.

Taylor, Frederick. *The Principles of Scientific Management.* New York: Harper and Brothers, 1911.

Weschler, Irving R. "The Leader Looks at Creativity." *Looking Into Leadership Monographs.* Washington, D.C.: Leadership Resources, Inc., 1961.

SELECTED REFERENCES

Anderson, Lester W. "The Management Team and Negotiations." *NASSP Bulletin* 53 (October 1969): 106-115.

Argyris, Chris. *Integrating the Individual and the Organization.* New York: John Wiley and Sons, Inc., 1964.

Argyris, Chris. *Interpersonal Competence and Organizational Effectiveness.* Homewood, Illinois: Irwin-Dorsey, 1962.

Atwell, Charles A., and Watkins, J. Foster. "New Directions for Administration: But for Different Reasons." *Junior College Journal* 41 (February 1971): 17-19.

Beckett, John. *Management Dynamics: The New Synthesis.* New York: McGraw-Hill Book Company, 1971.

Behavioral Science and Educational Administration. National Society for the Study of Education, 63d Yearbook, Part II, Chicago, Illinois: University of Chicago Press, 1964.

Bennis, Warren G. *Organization Development: Its Nature, Origin, and Prospects.* Reading, Massachusetts: Addison-Wesley Publishing Company, 1969.

Blau, Peter M., and Scott, W. Richard. *Formal Organizations.* San Francisco, California: Chandler Publishing Company, 1962.

Campbell, Roald F. et. al. *The Organization and Control of American Schools.* Columbus, Ohio: Charles E. Merrill Publishing Company, 1970.

Campbell, Roald F.; Corbally, John E. Jr.; and Ramseyer, John A. *Introduction to Educational Administration.* Boston: Allyn and Bacon, Inc., 1966.

Conwell, Anderson A., ed. *Administrative Team Leadership in Concept and Practice.* Athens, Georgia: Institute of Higher Education, University of Georgia, 1966.

Diener, Thomas J. "A Team Approach to Academic Administration." *School and Society* 99 (December 1971): 504-507.

Deneen, James R. "Role of Administrators." *The Encyclopedia of Education* 1:81-88. New York: The Macmillan Company and The Free Press, 1971.

Drewry, Galen N., and Diener, Thomas J. *Effective Academic Administrative: A Team Approach.* Athens, Georgia: Institute of Higher Education, University of Georgia, 1969.

Drucker, Peter F. *The Effective Executive.* New York: Harper and Row, 1966.

Drucker, Peter F. *The Practice of Management.* New York: Harper and Row, 1954.

Fiedler, Fred E. *Leader Attitudes and Group Effectiveness.* Urbana: University of Illinois Press, 1958.

Feltner, Bill D., ed. *The Administrative Team: Relationships to Internal and External Groups.* Athens, Georgia: Institute of Higher Education, University of Georgia, 1968.

Feltner, Bill D., ed. *College Administration: Concepts and Techniques.* Athens, Georgia: Institute of Higher Education, University of Georgia, 1971.

Fensch, Edwin A., and Wilson, Robert E. *The Superintendency Team.* Columbus, Ohio: Charles E. Merrill Books, Inc., 1964.

Griffiths, Daniel E. *The School Superintendent.* New York: The Center for Applied Research in Education, Inc., 1966.

Heald, James E., and Moore, Samuel A., II. *The Teacher and Administrative Relationships in School Systems.* New York: The Macmillan Company, 1968.

Hicks, Herbert G. *The Management of Organizations.* New York: McGraw-Hill Book Company, 1967.

Jackson, William H., and Snyder, Eugene E. "Team Administration at Santa Barbara High School." *Journal of Secondary Education* 46 (March 1971): 127-130.

Miller, Ben. *Managing Innovation for Growth and Profit.* Homewood, Illinois: Dow Jones-Irwin, Inc., 1970.

Morphet, Edgar L.; Johns, Roe L.; and Reller, Theodore L. *Educational Organization and Administration: Concepts, Practices, and Issues.* Englewood Cliffs, New Jersey: Prentice-Hall, Inc., 1967.

Ready, R. K. *The Administrator's Job: Issues and Dilemmas.* New York: McGraw-Hill Book Company, 1967.

Richardson, Richard C., Jr. "Needed: New Directions in Administration." *Junior College Journal* 40 (March 1970): 16-22.

Townsend, Robert. *Up the Organization.* Greenwich, Connecticut: Fawcett Publications, Inc., 1971.

OSCAR T. JARVIS
RICHARD W. BURNS

The Administrator and the Changing School Curriculum

Any discussion for the administrator of curricular needs, problems, programs or plans could fill several volumes, and so it is with reservation that one attempts to select from the complex of interrelationships which exert pressure on implementing or restricting curricula those topics which are most worthy of treatment. Along with the immensity of the selection task goes our personal biases and, as a result, we have chosen a limited number of needs and programs for discussion in this chapter. These needs and programs we believe are examples reflecting in some small way the "complex" problem which falls under the rubric of *the administrator and the changing school curriculum*. The examples selected and presented in this chapter in no way indicate that we are not aware of, or that we are not concerned with, those factors which others might consider important. We are not, therefore, treating a broad array of educational needs related, for example, to crime prevention, moral or character development, wise use of leisure time, and open-space classrooms. Perhaps these and other needs are equal to or even of more importance than those selected for discussion here. In this same sense, we are not discussing a host of special programs designed to implement curricula to meet all educational needs.

Concomitant with curricular needs and programs are problems and pressures, some of which are referred to briefly here, but none of which are treated adequately for "understanding." Again, we are aware of the many variables such as state laws requiring courses, local board of education policies, lack of money, lack of adequately trained personnel, outdated facilities, geographic isolation, and other factors which impinge on curricular and administrative decisions. In the few pages of this chapter, we are considering selected needs and problems which illustrate the nature of: (1) the interrelatedness of curricular factors, (2) financial concerns, (3) age level concerns, (4) special program concerns, (5) the need for innovative planning, (6) social demands, (7) cultural problem demands, (8) non-institutional programs, (9) weaknesses and strengths of existing programs, and (10) imperative program changes. May we suggest then, that the remainder of this chapter be read from the point of view that—*these are examples only* which reflect the authors' views. The concerns we address ourselves to are merely typical of the pressing contemporary problems associated with the issue of *the administrator and the changing school curriculum*.

This chapter, therefore, is mainly concerned with four major issues. These issues constitute the taxonomy of rubrics including: (1) contemporary curricular problems, (2) high priority curricular programs, (3) general curricular guidelines, and (4) future challenges for curriculum change and administration. A discussion of each follows.

CONTEMPORARY CURRICULAR PROBLEMS

We have selected four major curricular problems for discussion in this section. These problems include: (1) the welfare crises, (2) the drug culture and students, (3) race and ethnic relationships, and (4) environmental education. Now, let us examine each of these issues.

The Welfare Crises

Perhaps the dramatic way to indicate that the "poor are always with us" is to describe the number among us who are poor. The problem with this is: What number and by who's definition? Estimates of the poverty population in the United States range from twenty to sixty million depending on the criteria used. An apparently reliable figure is the number of welfare recipients. A June, 1971, report showed 16.4 million people were welfare recipients which was more than twice the number in June, 1962 (7.2 million).[1] The number varies radically from city to city and state to state. Generally, there are more whites than non-whites while proportionately the non-whites are in the greatest need. Rural areas demonstrate more poverty than urban, yet New York City with 1.2 million on the welfare rolls is faced with a tremendous concentration of cases.[2] It is not the cost in dollars that is the major poverty problem, although one cannot deny the financial burden, but rather the personal or psychological damage that poverty imposes on individuals. Sociological studies of children and

1. *U.S. News and World Report,* 14 February 1972, p. 22.
2. *Ibid.*

adults reared in poverty use such terms as hungry, depressed, hopeless, anxious, tragic, alienated, withdrawn, insecure, ashamed and dependent; all of which indicate some degree of emotional trauma. Plant early concluded that young children living in sub-standard environments were significantly less secure than children raised in more advantageous circumstances.[3]

Some authorities speak of the "poverty culture," a term appropriate to those born into and reared in disadvantaged environments. The psychologically damaging effects of temporary unemployment are naturally less severe and less evident in special behaviors than are the damages to the poor of the "culture." Some of these special behaviors such as violence, riots, and strikes, frequently are interrelated with racial, drug, and ethnic group issues.

It is beyond the scope of this presentation to describe all of the implications of poverty, but it does appear that the three factors mentioned, namely psychological damage, violence, and interrelatedness of causes and effects, all have serious educational, especially curricular, implications. Schooling, and particularly career and job training, offers one promising approach to alleviating the problems associated with those who we would class as the "have nots." Specifically, we should reasonably expect education of the poor, for example, to: (1) lower divorce rates, (2) improve health standards, (3) increase longevity of life, (4) reduce infant mortality, (5) decrease incidence of some communicable diseases, (6) decrease incidence of drug addiction, (7) decrease incidence of alcoholism, and (8) lower the rate of unwanted pregnancies. Granted that these specifics are descriptive social conditions, their close association with economic status has been shamefully exposed and routinely over-documented. Schools, at all levels, normally have not been responsive enough to the "problems of the poor" although some recent and current programs including Headstart, Job Corps, migrant worker training, school lunch, and scholarships and loans have been nominally successful. Much more money and effort to implement realistic and long-range programs are pressing needs.

In the immediate future, an attack needs to be made to reduce the number of unemployed heads of families in the nation which, in February, 1972, was reported to be 3.2 per cent.[4] In the long run, adult and continuous educational programs need to become an integral part of our educational system so that the human resources of our country are constantly

---

3. James Plant, *Personality and the Cultural Patterns* (New York: The Commonwealth Fund, 1937), p. 205.

4. *U.S. News and World Report,* 21 February 1972, p. 33.

being improved. The changing employment picture needs to be studied and predicted so that appropriate educational adjustments can be effected. All of this will help reduce the severity of the welfare problem. The schools alone cannot solve the problems of the "poverty culture" but they can play a major role in resolving them.

The Drug Culture and Students

As surely as there is a culture of poverty, there is a culture centered around the use of drugs. The drug problem is not confined to drug addicts, nor to young teenagers, nor to high school and college campuses. The use of drugs in our society is obviously beneficial in a controlled medical context and just as obviously abused in an uncontrolled escape sense. The hypochondriacs, as well as those commonly called "drug users," frequently employ some type of internally consumed chemical to effect a *release giving* physiological state. It is the close association of drug use to social maladies that gives rise to our drug problem.

Although the use of drugs to relieve pain, reduce tension and escape reality is very ancient in origin, its current impact is intense because of increased numbers of persons using drugs, urban concentrations of people, the easy access everyone has to them, and the wide variety of chemical compounds available. Furthermore, the problem is intensified by the fact that the use or mere possession of some drugs is illegal. The drug problem then has two major features; one the physiological effects on the individual, and the other the social antagonisms which result from drug abuse.

Basically, the consumption of drugs is not the major problem to be dealt with but rather the underlying personal and social cause which leads to drug abuse. Drug use is more of a symptom of personal-social ills than an ill in itself. Poverty, unproductivity, alienation, family disunity, dissatisfaction with self, school, and many other conditions which may be casual in nature seem to underlie the current widespread abuse of drugs. Today the use of drugs involves many of our institutions including the church, the family, the military, the government, and our schools. Additionally, the use of drugs is now widespread, involving all ages and all classes of people whereas historically this was not the case.

The drug addict may get the attention and the headlines because of his overt behaviors, many of which are dramatic and involve others as in robberies and other crimes, but our public schools are involved with many times the number of occasional users than they are with addicts. Recent indications are that the number of college students admitting to using drugs is steadily rising. For example, whereas in 1967 about 5 per cent said they had tried using marijuana, just four years later about 52 per cent responded the same. In the same period the sampling of hallucinogens had increased an almost unbelievable 1800 per cent.[5] The addict, of course, is a special problem and his plight is not without educational overtones. Addicts are mainly city dwellers because that is where the action is. In general, drugs, especially the narcotics, are almost entirely associated with the urban marketplace as they must be sold, there must be large numbers of buyers, and the expenses associated with addiction must be met.

The education or re-education of the drug addict centers around what is normally called rehabilitation. Although public schools are rarely involved in rehabilitating, the concept is worth mentioning for the insight it provides in changing human behavior. Rehabilitation is not, of course, confined to heroin addicts nor to alcoholics but in a larger social sense to reforming anyone with intense antisocial behavior. In general, rehabilitation demands two types of intense effort: (1) total immersion of the individual in a constructive environment where all facets of his behavior receive attention, and (2) total personal commitment of the individual to the direction of his reform. The rehabilitation concept has implications for our schools in dealing with intense deviant behavior of which drug use is one.

To return to the occasional drug user and schools, the present problem involves the increasing number of upper grade, high school and college students who have tried drugs. Student use of drugs can hardly be blamed on poverty or ghetto type environmental conditions. The motivation for student use of L.S.D., alcohol, marijuana, barbiturates, amphetamines, glue sniffing, sleeping pills, tranquilizers and other agents appears to be deeply associated with some form of alienation or rebellion. In no small sense must one ignore the school as a subculture—a world within a world where social pressures operate to initiate and sustain drug use as a means of identification with a peer group. In this case, the school youth who have, for whatever reasons, developed dissatisfaction with our society, its values, constraints and goals, have in turn developed their own subculture. Frequently, drug use has been a feature of this subculture and it probably serves several roles including, release from anxiety, a means of protest, a means of peer identification, a group ritual, a means for lowering inhibitions, and to some, a means of mind-expansion.[6]

Probably the majority of students try or use drugs as a form of protest. They are sensitive to and trying to tell society something about its adher-

---

5. *Newsweek,* 21 February 1972, p. 80.
6. The use of drugs to extend mental activities is another educational implication of drugs, the consideration of which is beyond the scope of this chapter. For an explanation of this concept see T. Leary, R. Metzner, and R. Alpert, *The Psychedelic Experience: A Manual Based Upon the Tibetan Book of the Dead* (New Hyde Park, New York: University Books, 1966).

ence to values associated with power-war-peace, environmental abuse, hypocrisy, discrimination, and other social phenomena which the students perceive as incongruent with their perception of ideals. Where failure and non-productivity may be offered as an explanation for drug use in disadvantaged environments, these reasons hardly apply to the majority of student users of drugs who may be high-ranking learners from middle and upper class homes. With the majority of students being only occasional users of drugs and the drugs most often used not being narcotics, it is difficult to prescribe a medical solution, a rehabilitation solution, or a punishment solution. In fact, the most obvious solution to the problem appears to be to *effect social change*.

The school's role in social change should be one of interpretation, enlightenment, and leadership. To the degree that outdated curricula, unnecessary prerequisite courses, non-relevant topics and information, irrelevant behavior codes, unrealistic counseling practices, unreasonable censorship of software, social restrictions, social discrimination, and lack of freedom in what and how to learn are characteristic of educational practices, then to that degree the school is abetting the problem of drug use!

What can schools do to diffuse the pressures which initiate or sustain drug use? Not enough is known about the student use of drugs and how to solve the problem to suggest "a solution." However, it appears logical that schools could effect many changes which would be positive in a large social sense and which would undoubtedly help to minimize drug abuse problems. Specifically, schools can:

1. Remove restrictive and non-relevant dress and conduct codes.
2. Refrain from identifying and grouping learners on the basis of socially deviant behavior which encourages subgroup formation of classes of deviants.
3. Allow students to have a greater voice in school affairs.
4. Allow greater freedom in selection or creation of courses and course content.
5. Allow greater freedom in selection or creation of learning modes and styles.
6. Revise curricula to increase the number and quality of youth-relevant and socially-relevant topics.
7. Provide for greater freedom of student self-expression.
8. Provide closer and more functional opportunities to participate in real-life community activities.

9. Provide an increase in the number and quality of creative opportunities.

Drug education must be mentioned as playing a role in helping to solve the school drug problem. Much effort and monies have been expended in this direction, but there is inconclusive or no evidence as to the effect of such materials and programs to date. Personal interviews of drug users have indicated that materials, talks, and programs which even in a small measure distort, preach, propagandize or in any way deviate from the truth are without value. The most effective programs and materials appear to be those which treat drug information and drug taking from a completely objective point-of-view. Young people will listen and *listen to reason* if they feel they are not being "put on."

Race and Ethnic Relationship

The United States, as a melting pot, has been an obvious setting for problems associated with racial groups involving such attitudes as prejudice, discrimination, rejection, and intolerance. When racial problems began or when they reached national proportions is hard to say—the overriding fact is that they have seldom, if ever, been worse than in the last ten years. Essentially, this problem also extends to religions and nationality background minority groups. Of concern here, are those problems which are current and likely to continue into the immediate future. On this basis, because of numbers of persons involved as well as social-economic reasons, there are Negro, Puerto Rican, Mexican-American, and American Indian problems besetting our culture and thereby our school system.

Because the term Negro is a specific racial term as contrasted to the use of the designation "black," in this context of racial and ethnic consideration, the term Negro will be used. It is recognized that black persons themselves wish distinctions to be made between Negro and black but since there are a variety of such connotations they are beyond the scope of this short discussion to deal with effectively. For this same reason, Mexican-American will be used instead of the term Chicano, a term more preferable in the Southwest.

The Negro and Puerto Rican problems are essentially, although not exclusively, urban in nature, while the Mexican-American problem is urban and rural, and the Indian problem is more rural than urban. Therefore, many but not all racial-ethnic difficulties are associated with urban and inner city education. That this is true is apparent in such cities as Baltimore, Chicago, Cleveland, Detroit, Philadelphia, St. Louis, and

Washington, D.C., where presently, or by 1980, the estimated non-white population is, or will be, in the majority. Any discussion of racial-ethnic problems, is, by necessity, intrinsically related to minority, urban, poverty, and cultural disadvantaged concerns and any consideration of racial-ethnic difficulties must be made in this larger social context.

Racial-ethnic concerns in our society involve prejudice; lack of educational, social, and economic opportunities; discrimination in housing, justice, employment, and promotion; lack of political power; lack of respect; and lack of recreational opportunities. Schools cannot, of course, directly provide the answers to all of these problems but there is much they can do.

First, schools must provide maximum facilities for educational opportunity. This means more than merely a school to attend. Opportunity also implies the availability and use of all the facilities and tools used in promoting learning. The United States social and economic structure and well-being demand that *all peoples* must be educated to their full potential. For any society to maintain a large majority in poverty or lower social status is costly in human resources, money, time and effort. Educational opportunity should not be provided, except perhaps as short term or emergency measures, in special programs designed for labeled minority groups such as poverty programs and compensatory education. This type of program intensifies the distinctions between groups already present and such programs are not always welcome by the minorities involved.[7] At best, these programs are usually temporary, often experimental in nature and they generally involve only a fraction of the minority group defined as eligible to participate. They are like the proverbial "drop in a bucket." When educational opportunity is provided it cannot by definition be through one or more special (different) programs.

No easy answer to educational opportunity is likely to be found, since education involves more subtle factors than can be provided by sheer dollar expenditures or a million good intentions. Very briefly, the problems referred to here involve known and perhaps unknown factors which interact in a system through or by which one succeeds. Success for the white child is one thing because he lives in and belongs to the system. Such is not the case for the non-white. Separate, but equal schools, neighborhood schools, zoning laws, closed suburbs, economic discrimination, lack of social integration, linguistic difficulties, and a host of other factors all effect not only success but the kind of education one receives. Learning is not limited to schools and schools can only operate in light of all the

---

7. Estelle Fuchs, *Pickets at the Gates* (New York: The Free Press, 1966).

constraints imposed by value and power structures. This is not offered as an excuse or reason for schools to remain unchanged or indifferent to the role they must play; but rather, as an explanation of reality which appears to be a slow evolution in the right direction.

Second, the curricular content of school programs needs to be closely examined and revised in the light of relevancy to minority rights, values, and heritage. Although some progress in this direction has been effected, it was only a few years ago that one could "go through school" and never discover, with perhaps one or two exceptions, that the Negro, the Indian, the Mexican-American, or the Puerto Rican minority groups had contributed anything to American culture: art, politics, science, music, mathematics, or whatever. Notably, however, one could erroneously learn that Columbus discovered America.

Third, grouping for instruction along lines that promotes *de facto segregation* must end. To group a bi-racial school's learners with slow and rapid learning criteria is to group them racially. There is no conclusive evidence that slow learners learn best when with others like themselves. Other insidious groupings involve the assigning or counseling of learners into special educational tracks or classes. This type of assignment frequently locates the minority learner in classes which in effect locks him into a disadvantageous educational position. Being assigned to related math may in effect assign the learner to a full schedule of classes of poorer learners. Or, the minority student may be counseled into vocational classes which train him only for a low-level occupation.

Fourth, the utilization of norm-reference tests (such as subject achievement, I.Q., or diagnostic) to group students for instruction also has deleterious *status quo* effects. Minority students, on the basis of test results, will again find themselves together with the resulting constraint of being unable to interact in meaningful patterns with the majority group. Social interaction is one feasible way for minority students to understand and acquire affective behaviors necessary for acceptance and belonging to the culture in which they will spend their lives.

There are many other school problems associated with racial-ethnic educational success. Again, this brief discussion cannot deal with them all. However, it should be recognized that early compensatory training, equal physical facilities, increased verbal facility, early childhood training, equal social opportunities, increased expectation, a hunt for talent and other such potential school-based activities can all contribute in some measure to reducing the present handicaps under which minority students operate in trying to learn.

Environmental Education

Strange as it must seem to man himself, the biggest question he faces today is *his own survival*. Homo sapiens and his precursors have for millennia been increasingly taking from the environment with no thought for the future. Even 100 years ago, it was hardly conceivable that we would ever run out of water or wood or wildlife—now we have little choice and maybe—just maybe—a little time. It is not much of a choice if one can merely decide to survive or not to survive. It would be much better if we could have many options on how to survive in harmony with our environment. It could be too late, at least for many of the world's population, to think in terms of surviving with a high standard of living.

Thirty years ago man was worried about nuclear energy and its threat to survival. Now, thirty years later, that threat, although still unresolved, pales when compared to the threat posed by the sheer numbers of mankind. Given our current rate of population increase, within thirty more years the population of the world will be twice as great as at present and there are not enough resources to go around now! With about seventy per cent of our three and one-half billion people undernourished in 1972, how are we to feed three and one-half billion more within thirty years? The problem of food alone, staggers the imagination.

The one problem of population is sufficient to justify the need for *environmental education* and yet, the number of people is merely one facet of a many sided problem involving the earth's natural resources—their use and abuse. We have data on stream pollution—perhaps not enough. We have doomsday prophesies—perhaps too many. What we don't have is *a national and international commitment to do something about getting the world's population and its needs in harmony with the world's resources.* Merely because President Nixon on February 8, 1972, in a message to Congress appears to have committed monies to the restoration and protection of the environment of the United States, we must not assume that the job is done. Cleaning up the contaminants that have already poisoned our streams, lands, air, and oceans, is just a start, not the solution to the problem. Until every citizen has a personal commitment to restoring and maintaining a complete balance of natural and man-made resources, we are falling short of what needs to be done.

This is where education enters the scene. There is little probability that people will become aware of and concerned enough to act relative to all the problems connected with resources, their conservation and use, unless they are educated about the problems. We should have no generation gap concerning environmental issues for in this case young and old alike must face common consequences, with or without action. Furthermore, we cannot talk solely about United States citizens because each person on the earth has a personal health, monetary, esthetic, and emotional stake in the world's resources. What happens in Kobe, Japan, and Nairobi, Kenya, is of concern to John in Winnebago, Iowa, or to Karl in Berlin, Germany.

It is not clear to scientists nor to educators exactly what needs to be done but whatever it is it must be a part of "total reform." We obviously need awareness, concern, and action but the latter cannot wait while we all get the facts. Many environmental needs must be acted on now. However, schools can certainly develop awareness and concern. It is difficult to think of a single school subject that could not and should not deal with environmental education. There certainly is no lack of topics to study nor problems to analyze. Moving from awareness to concern means going beyond cognitive levels of learning to affective levels. It is in the area of affective learning that the real problems of environmental education lie. How does one teach children and young adults to *want* to conserve resources, to *value* a disappearing species in Africa, to *involve* him or herself with the problems of pollution, population control, land use, and the one hundred and one other environmental concerns? We have a long way to go before we can do an effective job in this area. The most viable approach we can suggest is to structure environmental education so that it becomes a *doing* type of learning rather than a *knowing about* type of learning.

We referred earlier to total reform, and it is in this sense that the child should learn. *By and through doing,* environmental education should place the individual in a *total immersion* type of learning situation where the learner is actively involved in his own community with real environmental problems. The community must serve as a living laboratory for environmental studies. Personal commitment can, in our opinion, only be attained by permitting each learner to become totally involved with some real problem of his choice.

Additionally, learners must develop attitudes and skills related to independent thinking and decision making on a wide variety of environmental problems. They should be encouraged to develop original plans and strategies for dealing with problems and issues at the community, state, national, and world levels. Young children should come to feel that they can do something constructive and that environmental problems are not just the concern of some remote governmental agency.

It would be redundant at this point to comment on the many additional needs to implement sound programs of environmental education. We obviously need money, trained personnel, and proper legislation to get the job done. We have the doomsday prophecies—we have thousands of technical reports—we know what we need to do. Now, it is up to our schools to get on with the task of developing appropriate knowledges, understandings, strategies, and skills relating to our understanding and use of the world's resources. Each state, as a minimum, should develop a sound program for all public education levels and this is an immediate need.

HIGH PRIORITY CURRICULAR PROGRAMS

The administrator is besieged with a broad array of reputed sure-fire curricular programs today which will solve the educational needs of his constituency. Obviously, we cannot discuss all of these but we have selected five which we think are universally applicable and helpful. They include: (1) early childhood education, (2) education for exceptional children, (3) career education, (4) community college education, and (5) adult and continuing education. Let us examine each of these viable curriculum programs.

Early Childhood Education

Of all the needed programs in the public schools today, none is more crucial than the necessity of implementing viable nursery and kindergarten programs in all schools. The reason for this is very simple—through meaningful preschool programs, preventive measures can be taken to avoid problems that later develop such as learning disabilities and linguistic difficulties.

We now know that ages two to six are the most fruitful years in a person's life for learning.[8] We also know that the school offers instruction in *an elaborated language code*.[9] And, we know that, if we do not get the poverty culture children into meaningful nursery and kindergarten programs, they will develop *restricted language codes* and inadequate conceptualization skills which will make learning for them extremely difficult, if not impossible, in the elementary, middle, and high schools.

The preschool program is not solely mandated for children of the poverty culture. To the contrary, since children between the ages of two and six have been found to learn readily, as Bloom points out, then it obviously follows that all children from more economically advantaged backgrounds would also profit from early educational stimulation as found in preschool programs.

How early in life should any child begin schooling in the form of preschool education? An Educational Policies Commission report recommends

---

8. See: Benjamin Bloom, *Stability and Change in Human Characteristics* (New York: John Wiley & Sons, 1964).

9. Basil Bernstien, "Elaborated and Restricted Codes: Their Origin and Some Consequences," paper presented at University of Chicago, March 6, 1964.

that all children be given preschooling experiences beginning at age four.[10] The University of Georgia Research and Development Center in Early Educational Stimulation worked with children over a five year period who began preschooling at age three. The best evidence indicates that children can profit from preschool training as early as age three.

Of course, three year olds can profit from educational stimulation for only about two and one-half to three hours per day. As a result, preschoolers from more advantaged type homes would likely profit from being at school daily for only a limited time interval. It is likely, however, that many children of the poverty culture would benefit from spending most, or all of the day, at school if the school operates a day care center in conjunction with the preschool program. The reason for this is obvious—disadvantaged children can be supervised within the day care center in such activities as play and rest and they can continue to have socializing experiences including the use of elaborated language codes with their peers. Moreover, the environment of the day care center is more conducive to learning and reinforcing learnings previously acquired than are the individual home environments of most of the children from the poverty culture.

Compensatory education programs in this nation, on the whole, have been nominally successful at best. The reason for this seems clear: Compensatory training involves *treatment* of instructional deficiencies. What is needed then is for the preschool education program to *prevent* instructional deficiencies (particularly among culturally disadvantaged children) so that a later need for compensatory training will be minimized. Also, one must not overlook the fact that preschool programs provide salutary outcomes for the culturally advantaged children as well.

The cost of adding nursery and kindergarten training programs have retarded their establishment in many areas of our nation. This no longer is an issue. We can no longer afford not to establish nursery and kindergarten programs. This statement is obviously justified as per the information given in the foregoing paragraphs of this section.

Education for Exceptional Children

Conservatively, one out of every eight children in the public school is exceptional in some regard. Exceptional children are those students who are enrolled in the school and differ significantly from so-called "average" children to such an extent that instructional modifications must be made

---

10. Educational Policies Commission, *Universal Opportunity for Early Childhood Education* (Washington, D.C.: National Education Association and the American Association of School Administrators, 1966), pp. 3-12.

for them in the regular classrooms, in special classes, or in combination arrangements. Some of the more common types of exceptionality include: (1) intellectually exceptional, (2) sensorily handicapped, (3) orthopedically and speech handicapped, (4) special health handicapped, and (5) emotionally and socially handicapped.[11]

Presently, many authorities in the nation believe that, wherever possible, exceptional children should be placed in regular classrooms for instruction. This practice seems to be acceptable in teaching many types of exceptional children, i.e., speech handicapped, gifted, partially seeing, hard-of-hearing, and most orthopedically handicapped children. Whenever possible the emotionally disturbed and socially maladjusted children need to be placed in regular classrooms so that they can profit from the "modeling" effect of "normal" children. Many schools follow the procedure of even placing EMR children in regular classrooms for some, if not all, of their instruction. Advocates of integrating exceptional children into the regular classroom situation emphasize that a more socially cohesive unit with more salutary pupil outcomes result. Of course, when these types of exceptional children are placed in the regular classroom, specialists, as resource people, should be made available to work with such children and their teachers as necessary. Some of these specialists might include a speech therapist and a hearing therapist.

It would appear that the education of most exceptional children can proceed effectively in the regular classroom on the whole if adequate resource people, special materials and equipment, and minor adjustments in the facility (such as ramps and rails for the orthopedically handicapped) are provided. In fact, one can seriously question the efficacy of providing special classes with their concomitant increased costs which result from decreased class enrollments and more highly trained teachers. As Rossmiller points out, the research studies have failed to show that special classes produce superior results when compared with the practice of placing exceptional children in regular classrooms, at least, in terms of improved academic performance.[12]

Over the next few years, it appears that more schools will attend to the instructional needs of exceptional children in regular classrooms than in special classes. The increased money that would have been used to

11. Oscar T. Jarvis and Haskin R. Pounds, *Organizing, Supervising, and Administering the Elementary School* (West Nyack, N.Y.: Parker Publishing Company, 1969), pp. 111-112.

12. Richard A. Rossmiller, "Dimensions of Need for Educational Programs for Exceptional Children," in *Dimensions of Educational Need*, ed. Roe L. Johns, Kern Alexander, and Richard Rossmiller (Gainesville, Florida: National Educational Finance Project, 1969), pp. 67-68.

create, staff, and equip many special classes will be used to augment educational opportunities for exceptional children in regular classrooms by diverting such funds for technical and consultative assistance, special equipment, and relevant materials for use by the regular classroom teachers.

Career Education

A prior discussion of "The Welfare Crises" immediately brings to mind a crying need for increased emphasis on vocational education. It is not to be inferred that career or job training alone will eradicate poverty or that only vocational courses are related to the problems of the poor. In fact, *going to school* and considering *all courses,* given our *present system, with vocational training* would not be the solution to the total problem. A poor child, entering first grade has the right to a free education and the right to an equal opportunity but, at present, they are predictively headed for failure for many reasons.[13] In the United States our typical schools reward affluence, cleanliness, good manners, respect, cooperation, hard work, and Puritan values at a level not generally achieved by children from disadvantaged homes. However, there is a need for vocational education as one part of a revised system of education implementing the rights of a free chance for an education. Our society should provide the *means* as well as the *opportunity.* Then, as one part of a total, revised program, vocational education needs to be offered and so structured that it becomes an effective part of the means whereby individuals prepare for full participation in our culture.

During 1971 when the United States unemployment rate was running near 6 per cent it was often observed that it really was not the lack of job opportunities causing the problem, but rather lack of persons with the requisite skills to fill the vacancies available. This type of observation is difficult, if not impossible, to document, but it has some ring of truth to it. However, specific job training is not the total answer. In some cases, job training may be necessary as a temporary, immediate or a stop gap measure to alleviate a crisis or dangerous trend. In the long run, our cultural future can best be served by *career education.* Job training is specific, limited and generally met by a few courses or a short period of training. As the answer to employment, increased earning power, and personal happiness, it falls short of what is needed. If jobs change or disappear from the marketplace, as they frequently do, we generally record more statis-

---

13. Sheldon and Eleanor Glueck, "The Home, The School, and Delinquency," *Harvard Educational Review,* 23 (1953).

tics for unemployed data reports. *Career education,* on the other hand, is longer, more comprehensive, more flexible and personally oriented.

The best way to comprehend career-oriented education is to list some of its many aspects and then the contrast with job training will be readily apparent. *Career education* should contain all of these features:

1. Start as early as the child enters school
2. Provide real contact with career people
3. Provide for setting personal career goals
4. Offer the opportunity to explore a variety of areas of work
5. Provide up-to-date information about all facets of occupation including salary, requisite skills, educational requirements, projected opportunities, fringe benefits, . . .
6. Practical work experience
7. On-the-job training
8. Placement service

It is not to be construed that *career education* stops at any educational level. Some careers may be fulfilled by only a high school level of education, others may start post high school but continue with advanced training while others may require a college education or graduate training. It is the goal of *career education* to produce all-around, well-trained, interested, dedicated and flexible candidates for employment in a broader area than merely "one job." It is *not* the intent of *career education* to force early career choices, impose adult bias or restrict in any way the personal preparation of an individual for his "life's work." Additionally, *career education* implies freedom of choice for exploring areas of interest, freedom of choice of some areas of study, freedom of choice for selecting some of the means for learning, the opportunity to receive academic credit for work experience, the opportunity to "earn while they learn," and the opportunity to "change one's mind" as often as desired.

To implement *career education* will require funds for acquiring the means although this may not be as expensive as one might suppose. In fact, *career education* can make extensive use of existing community resources at considerable saving to a school system. One often wonders why "getting an education" is, to some people, synonymous with "school building." In the same vein why is "education" synonymous with "trained (certificated) teachers"? For instance, might not a hospital with its varied staff (from building engineer to surgeon) be an ideal place and means to earn extensive money and credits toward a medical career?

Career education should be a complement to our present social-economic structure. Frequently, vocational education is separated from academic education; but in today's world neither academic nor vocational education, by itself, is sufficient to the task of turning out a literate individual who can effectively fulfill all the roles demanded by our society. Whereas, an individual's employment is probably the most single demanding human activity, other social-civic-personal responsibilities require more than merely vocational training. In this sense, *career education* is intrinsically a complex blend of basic, recreational, social, economic, and esthetic skills as well as salable skills for employment. In a real sense, the work that a person engages in is not separated from marriage, rearing children, vacations, civic participation, voting and the many other human activities associated with living. It is exactly to this interrelatedness of human behaviors that *career education* addresses itself.

Community College Education

It should be readily apparent by now that "getting an education" is not a simple affair nor is it merely "going to school." In the sense that one learns all the time—birth to grave—that many skills are needed by an individual to participate in our culture and that our cultural vitality demands members with thousands of different skills, it is no wonder that K-12 schools cannot fill the bill. There are many programs outside the K-12 dimensions, but it is in the post high school sphere at present that we find the least educational universality.

There are thousands of colleges and universities, public and private, but the total enrollees earning Bachelor's and first professional degrees in 1969-70 numbered only 784,000 as compared to 2,978,000 high school graduates the same year. Although 59.8 per cent of the high school graduates went on to college that year, the majority will not complete a degree.[14] There is an unmeasured but obviously great loss in human resources at the end of high school given the assumption that many skills do, in reality, require additional training for acquisition and mastery. Whereas, all states provide tax-supported education 1-12, very few states provide such opportunity beyond the high school.

The one universal education movement, and therefore the one with the greatest chance to become a successful reality, is the Junior and Community College movement. For our purpose here, we shall consider these two institutions, even though different, as representing a single movement (community schools) meeting post high school needs. There are several reasons why Junior and Community Colleges have been established, some of these are:

14. *Statistics of Trends in Education 1959-60 to 1979-80* (Washington, D.C.: National Center for Education Statistics, U.S. Department of Education, Office of Education, March 1971).

1. To provide an educated source of talent for entrance to senior colleges, perhaps at junior level.
2. To provide wider educational opportunity than is provided by regular colleges.
3. To provide skilled workers for an increasingly technological culture.
4. To provide skilled employees for special and drastically needed service areas (hotel, medical, areospace, etc.)
5. To provide continuing educational opportunities in areas not normally a part of regular college and university curricula.
6. To provide continuing educational opportunities for all community citizens—similar to adult education.
7. To provide post high school opportunities in more accessible places—nearer the student's home residence.

Not all J.C.'s and C.C.'s are meeting all of the educational needs as previously specified; some have programs that are general in nature, some offer specialized programs, some encourage curricular offerings tailored to community interests while others are almost entirely preparatory for senior college level instruction. The growth of this type of institution has been phenomenal in the past few years with over 500 new community colleges being established during the 1960's. In 1960, approximately one out of five post high school students began a thirteenth year of education in a community junior college and it was estimated that in 1970 the number would be approximately one out of three.[15]

With all of the growth and variety of J.C. and C.C. institutions, much remains to be done in extending community junior college post high school educational opportunities. In a few states, notably California, Texas, New York, Iowa and Florida, there are rather extensive community school programs, but in other states there are very limited opportunities for continuing one's education. Of late, the Federal Government and some national groups have contributed, through grants and political pressure, incentives for improving community college training but the burden of effort has been and remains at the community level. Community representatives have recognized their needs and frequently have done something about them, often with the profound realization that through education their community and our country will grow and develop.

It appears that there are several needs which the community school can fulfill which either are not being currently met by other institutions, or which can be more uniquely fulfilled by a well-conceived local institution. Most of these needs will be of an adult, vocational or career educa-

---

15. Edmund J. Gleazer, Jr., *This Is The Community College* (Boston: Houghton Mifflin Company, 1968).

tion nature. First, community colleges can operate to meet local educational needs other than those of the 18-21 year age range. Community schools can offer courses designed for all adults or for special groups of adults interested enough to generate and request such courses. Second, community schools can offer courses which meet special community needs for vocational skills determined to be in short supply. Third, such schools can maintain an open door policy for entrance thus making available educational opportunities to individuals who cannot meet entrance requirements to academically oriented colleges and universities. Fourth, community schools can compliment and often complete career education started in earlier school years when such career training does not require a four-year college program. Examples of this would be medical technology, legal secretary, law enforcement, mid-management executives, hotel and motel operation, X-ray technicians, and teaching aides. Fifth, these schools need to offer many non-credit (college level) courses yet the courses themselves should be of high quality. There may be community need for conversational Spanish, blueprint reading, photography, antique furniture restoration, or courses of a hundred other possible titles. Some of these non-credit courses may be of a hobby or recreational nature designed to assist people who are experiencing increasingly shorter work weeks and increasing time for leisure.

As there are needs, there are also some needn'ts—things Junior or Community colleges should not do. These recommendations appear logical:

1. They should not offer programs competing with those available in four year colleges and universities.
2. They should avoid the overemphasis on general and liberal arts areas frequently typical of the first two years of four year colleges.
3. They should avoid high, restrictive entrance requirements.
4. They should avoid high financial costs to enrollees.
5. They should avoid losing control over their own destiny—community schools should be governed and administered locally although tax support may be broadened.

Additional, potentially beneficial, features or policies of community colleges which should or could be incorporated into their operation involve such matters as:

1. Employing local talent not all of whom hold academic degrees. (A retired cabinet maker just might be the ideal instructor for a course in cabinet making.)
2. Utilizing community resources already in existence as libraries, physical space, and on the job training sources.

3. Granting credit for work experience.
4. Establishing club sport programs instead of emphasizing varisty inter-school competition.
5. Extensive use of individual or self-instructional materials.
6. Offering opportunities for interaction between students and non-student community residents. Some return to the concept of the school being a community center would appear to have beneficial effects on both the school and the non-school community.

The community school movement holds great potential for providing needed educational opportunities for all types of adults, especially those not interested in college (4-year) programs, and in curricular areas not normally attended to by colleges and universities. This movement is growing, as well it should, and its potential has hardly been tapped.

Adult and Continuing Education

One of the best needs for adult education just arrived by mail this morning, and it said:

As for going to school with all those young people: A generation gap may exist, but today's young people are eager to close that gap. They are both intelligent and receptive. If you're willing to listen and discuss issues with them, you may find that some of your ideas will help them become wiser and more mature individuals. And some of their ideas may help you become wiser—and younger.[16]

The author of the above quotation, Paul L. Garner, was talking about adults going to a Junior College, and we have already mentioned the need for that type of Community School. There are many adult needs for education, some of which are or can be a part of the community type of school; but most communities need other continuing education classes which, at present, are not associated with a high school, community school or an institution of higher learning. For example, there are special high school programs (evening classes), YMCA classes, professional in-service training, correspondence schools, private schools, church sponsored learning activities and other similar activities. It is this lack of official (governmental) recognition and institutional support that is one primary need in adult and continuing education today.

A person should not confuse adult education with a vocational education, educational opportunities for the aged or literacy education. In fact, some examples of the best adult education is that associated with non-academic institutions, usually profit making organizations which regularly

---

16. Paul L. Garner, "College Is Also For You," *Modern Maturity*, Vol. 15, No. 1 (February-March 1972), p. 33.

train or re-train individuals, thus enabling the sponsoring organization to maintain a competitive position in our rapidly changing times. Technological innovations, foreign competition, pollution controls, anti-discrimination policies (race and sex) as well as other pressures all but require business and industry to implement pre- and in-service training programs. This pressure is automatically extending itself to governmental agencies by way of police training, law enforcement, fire fighting, and other types of special schools or classes. A recent example is afforded by current plans by the Food and Drug Administration to broaden its training program for its food inspectors and to offer like training to employees of private businesses which will enable producers to turn out better products. This type of adult education is more frequently termed *continuing education* and it accounts for untold increases in earning power, productivity, tax revenues from increased earnings, job satisfaction, and higher living standards. Some of this type of training is certainly best accomplished within the framework of the job environment; that is, at the job site rather than in a public school, college, or university. However, it is an interesting question to ask if some of the training couldn't be more efficiently handled by schools, and if so, why aren't the schools involved? It may be that schools have not been sensitive and responsive to societies' needs.

Examples of industry-college level and government-college level cooperation in educational ventures are not rare, but on the other hand, they are far from extensive. Institutions of higher learning do cooperate with local law enforcement agencies, insurance companies, and communication industries (notably AT&T), to name a few, in conducting workshops and training programs. Most of these programs (or classes) are closed sections (open only to hand-picked or subsidized enrollees) and one wonders if it might not be more academically sound to hold mixed classes. By this we mean, allow enrollments in such classes to come from the student body and the community at large as well as from the exclusive business, industry, or governmental agency. Would not social and intellectual interaction between mixed types of students enhance understanding and relationships between segments of our society who do not, at present, always see eye-to-eye? One wonders what behavioral changes of an affective nature might result from a class whose composition was one-half police officers and one-half college students, or in a class one-half middle level executives and one-half housewives.

One can't help but speculate on other problems which might be reduced in intensity through well conceived adult and continuing education programs. Might not the "generation gap" be less gaping; the drug problem reduced or the communication problem made less intense through innovative adult-youngster community learning programs? For example, the communication problem, by itself, is so far reaching as to be almost a disgrace to a country which prides itself on its schools. It recently has been reported that communication ability of less than eighth grade level causes one to be severely disadvantaged in our present culture.[17] Unless something drastic is undertaken, this problem will only become more intense for a number of reasons, e.g., we lack competency-based educational programs, our drop out rates continue high, and our society is growing more complex. The lack of communication skill and functional learning habits mitigates against any easy solutions to training or re-training of: (1) welfare cases, (2) those employed in unskilled labor jobs which are disappearing, (3) early school dropouts who wish to "return," (4) migrant workers, and (5) those functionally illiterate and another language as well as English.

What needs to be done? Mention has been made of some problems, but many facets of this complex area of adult and continuing education are beyond the scope of this discussion. High priority should be given, however, to:

1. Greater expenditure of funds (all sources) for new programs as well as extending existing ones.
2. Greater governmental (local, state and federal) involvement in establishing new programs to meet a variety of needs.
3. Greater cooperation between schools (all kinds), business and industry and the non-business community.
4. Bringing education to the people rather than expecting people to "go to school."
5. Regulation of private schools (including correspondence schools) to ensure they are meeting some type of accreditation standards.
6. Increased counseling services which will effectively reach the disadvantaged and alienated persons whose educational needs are the greatest.
7. Articulation problems, especially from high school onward, involving all segments of academic and non-academic learning entities.

In summary, tomorrow is always another day but the postponed problems have already caught us. No one knows when a crisis is a crisis so it is easy to say, without any real proof, that job training, self-improvement needs, and other such problems which can be solved by education are at the critical stage. However, one only needs to pick up a newspaper to read

---

17. Management Technology Inc., *A Comprehensive Plan for the Solution of Functionally Illiterate Problem, A Report on the Present, A Plan for the Future* (Washington, D.C.: Office of Education, March 1968).

about riots, welfare problems, migrant workers, unemployment statistics, geriatric needs, lack of medical services and a dozen other topics all of which can be partially resolved by intelligent application of resources presently available. We have learners wanting to learn, teachers looking for jobs, unemployed workers with time on their hands and school buildings vacant evenings and summers. Why can't these entities be brought together through adult and continuing educational programs?

GENERAL CURRICULAR GUIDELINES

In our introduction to this chapter we explained that to discuss all the factors relating to the administrator and the changing school curriculum would be far beyond the scope of the chapter. Therefore, we selected a few change needs and a few problems for discussion. However, we feel that some mention should be made of the total picture, especially those factors which must be considered in the light of present educational needs and trends. Certain general concern must be attended to in devising and administering curricula at all levels of schooling. It would be impossible to establish any hierarchy since all of the considerations mentioned in this section are important—none can be neglected.

1. *Curricula must provide for individualized opportunities to learn.* There is no "average" learner, no way a group can successfully be instructed by a method, a text, one set of materials, one topic, etc. The individual is not a constant in the learning process, rather, he is a variable.
2. *Curricula must provide for learner choices.* There must be alternatives offered or, perhaps in the future, freedom for the learner to choose *what to learn* and *how to learn.*
3. *Curricular content must be relevant.* Relevancy has two major implications; namely, relevant to the individual and his life, and relevant to the culture in general.
4. *Curricular content must hold more potential to learners than mere acquisition of information.* Requiring mastery of information of so-called academic curricula by all learners is truly an exercise in futility. Fulfilling the needs of learners should mean teaching processes, developing problem solving skills, attending to affective dimensions, and developing vocational skills in addition to the acquisition of some knowledge.
5. *Curricular content must address itself to societies' problems.* We cannot justify requiring of everyone the memorizing of Chaucer, dissect-

ing crayfish, analyzing *Tom Sawyer,* solving unrealistic physics problems (compute the density of a gold brick) and achieving similar traditional objectives at the expense of developing needed skills related to racial, ethnic, poverty, pollution, and other serious society concerns.

6. *Curricula must reflect local or community needs.* Because we are a heterogeneous society living in diverse geographic locations, we need some degree of flexibility in curricular designs. Local needs, from recreational skills (skiing in Vermont to rock hunting in Arizona) to vocational skills at all levels from pre-school to adult must be attended to if curricula are to be functional.

7. *Curricula must reflect the concept of special concerns for special groups.* General curricula or standard curricula will not serve the mentally handicapped, the physically handicapped, the socially deviant, the drug addict, the unemployed, the immigrant or other groups with special needs. To conserve and maximize the use of human resources, is to be assured that everyone does have the opportunity to learn.

8. *Curricular offerings must be implemented by appropriate means.* Educational opportunities must be brought to the learner and made available at a cost the learner can afford to bear, on a cost/effective basis to society, and in a non-aversive way. We cannot continue to blame learners for program deficiencies, offer sub-standard opportunities to the poor, and otherwise administer curricular programs in ways which punish or exclude anyone eligible to participate.

9. *Curricular goals and objectives must be specified.* We are long past the time when we can defend curricular requirements only by saying "it is good for you" or "you will need it later." We cannot afford to cover up what we, as educators, are (or are not) doing by disclaiming responsibility under the guise of "we are educating the whole child" or "we are teaching life adjustment." Accountability is here or just around the corner.

10. *Curricula must prepare learners for change.* Flexibility, adaptability, innovativeness, and independence in the pursuit of knowledge should be emphasized in curricula because our society is rapidly requiring such traits as it becomes more complex, more technological and as the store of knowledge becomes greater. An educated person is no longer one who has X number of courses or spent Y time in a school.

In addition to these major guidelines, or perhaps pertaining to them, are dozens of more specific principles which should be considered in designing and implementing curricula. One example is afforded by the considerations which pertain to individualized opportunities to learn. Although individualized instruction in the minds of some is a method rather than a curricular feature in learning, there are certain curricular implications such as:

1. The learner should be able to select topics or create topics to learn.
2. The learner should be able to select materials or create materials enabling him to learn.
3. There should be means whereby learners can "test out" (pretest) topics they already know. This is perhaps more appropriately expressed as the learner being able to learn (acquire) only those behaviors that are not in his behavioral repertoire.
4. The learning materials should be properly sequenced to maximize learning efficiency.
5. Learning should proceed in small increments.
6. The curriculum should be designed to maintain the behaviors learners have already acquired.
7. There should be a variety of curricular materials available to the learner.
8. If learners forget, they should have the opportunity to review or relearn.
9. Learners experiencing difficulty in learning should have the opportunity for interacting with alternative materials and methods (recycling) or with curricular materials designed for remedial purposes.

Competency-based education also has curricular implications. This movement is gaining momentum and therefore will be more influential in the near future than at present. In competency-based education the curricular outcomes are treated as constants rather than variables as in our traditional programs. With the learner, time and methods varying, the competency criteria are specified in terminal behavioral objectives.[18] It is by means of objectives that educators and the public have finally "gotten a handle" on the outcomes of education. Objectives express end-products of learning in *measurable terms* and this, in turn, has introduced the notion of *accountability*. To date, educational effectiveness has been based largely on very indirect measures such as: (1) number of staff members,

---

18. Richard W. Burns, *New Approaches to Behavioral Objectives* (Dubuque, Iowa: Wm. C. Brown Company Publishers, 1972).

(2) training level or degrees held by teachers, (3) space, (4) library volumes, (5) teacher contact hours with learners, (6) funds expended, and (7) availability of other resources.

Cost accountability is easily arrived at given the assumption that learning can be effectively measured. All, or any segment of instruction, can be measured at the performance level and compared with the dollars expended in achieving it. The restrictions up to now have been in obtaining reliable and realistic measures of learning. With learning defined in behavioral terms, a new type of test has been developed called a criterion-referenced test. Such tests, frequently utilizing processes and products rather than verbal items, measure the degree of interaction between the learner and the learning strategy. From tests of this type, we can collect data concerning instructional effectiveness.

There are pitfalls in the cost effective/criterion-referenced system that are easy to stumble into, and apparently some educators already have. With patience, time, and more work, an acceptable job of accounting can be accomplished which will avoid at least the obvious problems. One problem to be solved is the exact specification of learning outcomes in behavioral or performance terms. One can quite easily write objectives, but this is no guarantee that they are good—that is, serve some useful purpose. To account for non-relevant performances would be a waste of time. Another problem involved in accountability is the tendency of some to place the responsibility in a fixed position; as, for instance, with the teacher. This type of restrictive accounting denies the fact that learning is a system involving learners, time, a strategy (learning environment) and a criterion measure. In reality, the teacher is only one factor in the strategy part of the system. To hold teachers strictly accountable for all learning without considering the other factors is indefensible. Might not the fault lie with the learners, the software, the hardware, the physical plant or the criterion measure? At any rate, the concept of accountability is an interesting one which undoubtedly will have a great deal of influence on curricular designs. It is too early to say what the effects will be.

FUTURE CHALLENGES AND CURRICULUM DEVELOPMENT AND ADMINISTRATION

There are many issues that need to be understood and resolved in effecting curriculum change and administration in the 1970's. A discussion of some of the more important issues in the authors' view follows. These issues include: (1) dispelling the myth of urbanism, (2) the need for relevance in education, (3) the need of a year-round school, (4) the need for a comprehensive school, and (5) systematizing curricular change.

Dispelling the Myth of Urbanism

Swanson, in 1970, pointed out that of 6,000 legally constituted cities in the United States, only five had a population of more than 1 million. Moreover, he stated that only 9.8 per cent of the total U.S. population lived in these five cities. Likewise, he showed that 16 per cent of the nation's population lived in municipalities having 500,000 or more residents. And he stated that while the proportion of the population in cities of less than 50,000 residents had increased by 50 per cent since 1920, the percentage of the population living in municipalities of over one-half million scarcely had increased at all. For that matter, Swanson has shown that the most prevalent urban place is characteristically a city having between 10,000 and 50,000 inhabitants.[19] It would appear then that we have created a rhetoric about the all-encompassing plague of the problems of the urban dweller that at best is somewhat fallacious. Perhaps it is time that the curriculum developer and administrator place these facts and rhetoric in proper perspective when considering curricular change.

Why then do many Americans avoid residency in large urban cities having 500,000 or more people? Why do many of them elect to live in smaller cities ranging in population from *10,000 to 50,000* inhabitants? The answer to these two questions will help us understand more clearly some of the curricular needs of students.

It would appear that Americans residing in the smaller cities where the most growth in terms of population has taken place since 1920, have former agrarian backgrounds but now wish to pursue more urban-oriented economic activities. At the same time, they wish to avoid the crowded conditions and the anonimity that the larger cities impose on them as inhabitants. Schools, in these smaller cities, are generally too small to independently achieve a high degree of educational efficiency and excellence, but they may participate in some sort of cooperative intermediate unit for shared services.[20] Generally, the school's adult clientele in these smaller cities range from semi-skilled laboring types through professional and business people. As a result, unemployment is normally low and social problems, such as drug use and crime, that are prevalent in the large urban cities are only nominal.

Conversely, the large municipalities of 500,000 and more inhabitants have had in recent years an in-migration of large numbers of unskilled manpower consisting, for example, of poor whites from Appalachia and poor blacks from the rural south. These people have migrated into the

---

19. Gordon I. Swanson, "The Myth of Urbanism," *Education Digest*, Vol. 26, No. 1 (September 1970), pp. 34-35.
20. *Ibid.*

inner city of the large urban areas in search of jobs having been displaced by technology in the mines of Appalachia and the farms of the south. At the same time, there has been an out-migration to suburbia of middle and upper class inhabitants from these large urban centers in an effort to escape the problems of the inner city which this influx of disenfranchised Americans has precipitated.

The foregoing discussion contrasting the smaller cities' problems and their implied educational needs with that of the few large urbanized municipalities, helps point out that there are marked differences in characteristics and needs of so-called "urban" settlements. What is common to both, however, is that we have escaped our agrarian past for now it has been estimated that five per cent of our population produces the foodstuffs and raw materials that feed and clothe the rest of us. Viewed from this perspective, there are common threads of educational need which confront all urban dwellers whether they are living in a municipality of 10,000 or 1,000,000 inhabitants. It is to these common needs that we address ourselves in the remaining sections of this chapter.

The Need for Relevance in Education

According to Venn, approximately one million students fail to graduate from the nation's high schools every year. Moreover, he states that of forty per cent of high school graduates who enter a four-year college each year, only one-half of them earn a baccalaureate degree. The question, in the light of these figures, might appropriately be asked: "How relevant is education for today's student?"[21]

Obviously, relevancy in curriculum for many, then, is not going to be found in college preparatory and vocational programs. As we discussed in earlier sections of this chapter, the search for and hope for relevancy resides in strengthening and rewarding the talents of each student as he seeks to identify with an adult role in society in which he can make a contribution. This means that some people without educational credentialing will have to be used within the public schools to purvey their expertise in perfecting the talents of students which today's curriculum too often does not attend to. As Venn has indicated, a curriculum needs to be established where, in addition to college prep programs, a program designed for career preparation and development is also available. Such a program would provide for the acquisition of basic educational skills in which theory and knowledge for a career are related, in addition to students attaining specific occupational skills at specified levels of competency for an adult role. Whether such training is acquired in high school, in college,

---

21. Grant Venn, *Man, Education and Manpower* (Washington, D.C.: The American Association of School Administrators, 1970), pp. 229-230.

or in a continuing education program, provisions should be made for students who pursue such a program to re-enter the education system at any time after their initial training has been completed for more learning and re-training.[22]

In the career preparation and development program, concomitant work experience should be required of students. Educational credit should be given for this experience and evaluated. Some justifiable reasons for providing work experience are: to explore about a career area, to learn about the attitudes, responsibility and cooperation expected, and to acquire the necessary skills for entering the occupational area force.[23] Of importance for the student, particularly for the student of the poverty culture, is that he be able to earn a wage, however small, while he learns in the career preparation and development program.

The Need for a Year-Round School

As we have stated, we are no longer an agrarian nation. There are no more western frontiers in our nation to conquer and, in concert with our rapid technological developments, there is basically no present need for unskilled and under-educated people. Moreover, it is unreasonable to assume that we should turn children out of school during the hot summer months in our urban centers where there are few, if any, meaningful pursuits for them to follow. This is particularly true for the disadvantaged learners who, more often than not, cannot afford the attrition in learnings that occur when they go through a three-month vacation period. Likewise, this nation can ill afford the luxury of closing its schools during the summer months when new buildings have to be constructed just to accommodate increased enrollments or when there are many unmet educational needs involving such areas as adult, career, and vocational education.

For these and other reasons, it seems evident that year-round schools are mandated.[24] That is, they are justified whether or not all children attend school for four quarters (ten or eleven months) annually or whether they simply attend for three quarters per year.

The Need for a Comprehensive School

Schools of the future will have different characteristics if their curricula are viably relevant. They will, in addition to college preparatory programs, be comprehensive by offering career preparation and develop-

22. *Ibid.*, 230.
23. *Ibid.*, 242.
24. For listings of reasons justifying the Year-Round School, see: *Ibid.*, pp. 112-114; Roe L. Johns, "The Extended School Year," in *Dimensions of Education Need*, ed. Roe L. Johns, Kern Alexander, and Richard Rossmiller (Gainesville, Florida: National Educational Finance Project, 1969), pp. 192-193.

ment programs. Characteristic of these schools will be occupational guidance programs, increased vocational training options, continuing education alternatives, pay incentives while students learn in work-study situations, closer alliances with social agencies and with business and industry as well as with the total adult society.[25] Perhaps the most important characteristic will be the new notion that *credentialing* and *graduation* do not obviate the need for continuing education—a concept that still persists among some who are concerned with curriculum change and administration.

Systematizing Curricular Change

Since this volume is mainly concerned with *A Systems Approach to Educational Administration,* it seems appropriate to conclude this chapter with a brief discussion of systematizing curricular change. Of course, any systems plan starts with a statement of *goals* which have their genesis in the social and cultural context. Once goals of change have been stated in generic terms, it then becomes necessary to state *specific objectives* which are descriptions of terminal behaviors necessary for the attainment of each goal. Having outlined the goals and specified the terminal objectives, it then becomes necessary to state the *inputs* into the systems plan. These inputs may include human, fiscal, and temporal resources.

Assuming one wished to change the curriculum of the elementary school to an individualized instructional plan, for example, the next sequential step following the marshalling of *inputs* (human, fiscal, and temporal resources) would be designing *strategies for implementation.* (Some examples of strategies for implementing individualized instruction might include modularized training programs of a self-pacing nature. Such modules might have the following characteristics: (1) a *rationale* as to why the pupil should attend to the module of work, (2) a statement of *terminal behaviors* to be learned, (3) a *pretest* to determine what behaviors in the module are not presently possessed by the learner, (4) an array of *instructional alternatives* that the learner might choose in mastering those behaviors not presently internalized in his behavioral repertoire, and (5) a *posttest* to check the acquisition of the terminal behaviors the student has been working on.) Implicit in the self-pacing modules of independent study is a systematic plan of *feedback.* That is, when recycling is necessary for the student to attain the prescribed terminal behaviors, the checks in terms of pre- and posttesting are built into the system to assure the student that he is proceeding in an orderly manner toward the attainment of all required, terminal behaviors.

25. Venn, *Man, Education and Manpower,* pp. 120-121.

The final step in the systems plan alluded to above is that of *evaluating* the outcomes. The evaluation should provide data which the administrator can use in determining whether or not the individualized instructional program change as previously outlined is salutary.

Selected References

Burns, Richard W., and Brooks, Gary D. *Curriculum Design in a Changing Society.* Englewood Cliffs, N. J.: Educational Technology Publications, 1970.

Burns, Richard W. *New Approaches to Behavioral Objectives.* Dubuque, Iowa: Wm. C. Brown Company Publishers, 1972.

Cook, Desmond L. *Educational Project Management.* Columbus, Ohio: Charles E. Merrill Publishing Company, 1971.

Guthrie, James W., and Wynn, Edward, eds. *New Models for American Education.* Englewood Cliffs, N. J.: Prentice-Hall, Inc., 1971.

Heidenreich, Richard R., ed. *Current Readings in Urban Education.* Arlington, Virginia: College Readings Inc., 1971.

Hymes, James L., Jr. *Teaching the Child Under Six.* Columbus, Ohio: Charles E. Merrill Publishing Company, 1968.

Johns, Roe L. et al., eds. *Dimensions of Educational Need.* Gainesville, Florida: National Educational Finance Project, 1969.

Johnson, Kenneth R. *Teaching the Culturally Disadvantaged.* Palo Alto, California: Science Research Associates, Inc., 1970.

Lessinger, Leon M. *Every Kid a Winner: Accountability in Education.* Palo Alto, California: Science Research Associates, Inc., 1970.

Olson, Paul A. et al., eds. *Education for 1984 and After.* Schiller Park, Illinois: Directorate of the Study Commission on Undergraduate Education and the Education of Teachers, 1971.

Rudman, Herbert C., and Richard L. Featherstone, eds. *Urban Schooling.* New York, N.Y.: Harcourt, Brace & World, Inc., 1968.

Venn, Grant. *Man, Education, and Manpower.* Washington, D.C.: The American Association of School Administrators, 1970.

Weinberg, Carl. *Education and Social Problems.* New York, N.Y.: The Free Press, 1971.

LARRY L. LESLIE

The Administrator and Instructional Innovations and Technology

INTRODUCTION

The purposes of this chapter are to present a rationale for educational innovation; to survey innovative teaching techniques, supporting technology, and innovations in administration; and to present the implications of these for the administrative process. The framework for this chapter will be the administration of elementary and secondary school innovations *within a systems approach.* Not all innovations will be presented; however, most major instructional developments of recent vintage or a representative of the various forms of innovation will be included.

Let us begin the rationale with a controversial assertion: Innovation is, in itself, "good;" without specification, change is by itself, "good."

"What nonsense!" many would say. The platitudes run deep in this regard, especially so, perhaps, among school administrators. "Let's not have change merely for the sake of change," is a remark often uttered around the schools in response to a new idea. It could be asserted with considerable accuracy that all too often either the conscious or unconscious motive behind such statements is the desire to maintain the status quo, for it is human nature to resist change. Change is a painful process. There is comfort in familiarity with settings and procedures. Innovation and change represent the unknown and the advantages they might bring over what presently exists and may be working smoothly, is uncertain. Why alter a perfectly acceptable system when change will absolutely require additional expenditures of time, energy and financial resources?

There are other, more personal, psychological reasons for resistance to change. The act of innovating has implications related to personal egos because the individual who agrees with the need for change necessarily acknowledges the shortcomings of present programs. This may not be a serious dilemma for the person new to an organization, although even such persons are probably committed to a certain ethos as a result of their training programs; but for the individual who has devoted a lifetime to a particular program, such an admission may be more than can be reasonably expected. This state of affairs may be difficult enough for senior personnel lacking major program responsibility, but for educational leaders recognition of the need for change may be next to impossible. For example, this attitude appears to be at the heart of the resistance to change away from the normal school concept in teacher preparing institutions, where faculty members share with administrators the authority to make curricular decisions.

Although these specific illustrations of change resistance among members of an organization are undoubtedly accurate, they fail to tell the whole story. How an organization resists change is clearly a much more complex matter than simple protection of self-interest, which is only one manifest form of such resistance.

The full theory of such resistance to change has been built by Donald A. Schon, who labels his theory, "dynamic conservatism." Schon points out that organizations do not resist change by sitting back placidly or apathetically, but they dynamically act to meet the challenges of change. They "fight to remain the same," says Schon.[a] Dynamic conservatism is much more than can be explained by "villainous machinations of vested interests," according to him; social systems, or organizations, hold a certain power over organizational members who see the connection between the survival and status of the organization and their own self-interests. To understand this connection and the power it has over individuals, it is necessary to realize that the organization is not only the source of each members livelihood, but also represents the major source of meaning to personal lives. Threats to the organization challenge the very values and meanings of life. Schon states that "the threat of organizational disruption plunges individuals into an uncertainty more intolerable than any damage to vested interest."[b] It is for these reasons that change can be brought about only by vigorous, if not, dynamic leadership—leadership dynamic enough to overcome the equally dynamic conservatism of the organization.

Let us return, then, to an earlier question, "Why change a perfectly acceptable system. . . ?" The answer is simply that the present system is *not* acceptable. In spite of some assertions appearing in the literature that

---

a. Donald A. Schon, *Beyond the Stable State* (London: Temple Smith, 1971)
b. *Ibid.*, p. 52.

elementary and secondary education has vastly improved over the past few years, the arguments of those who disagree (e.g., Charles Silberman, John Holt, William Glasser, Herbert Kohl) are more convincing. To these men, American education continues to be characterized by an endless duplication of stereotypic classrooms, each containing 30 to 40 students, all day every day—classrooms in which students almost universally are required to sit in neat rows of chairs, which may be bolted to the floor, but in any case, their arrangement is such as to require all eyes to the front so that the teacher may dominate all activities.

This appraisal of American education is reconciled with the contrary views expressed in much of the literature by considering the origin and circumstances of the latter. Most of this literature is written by education professors who rarely journey more than 30 miles from campus to make school visits and even then their purpose is very often to review some colleague's innovative project. (This statement is not intended critically; to venture further afield would serve little purpose.) Further, it is not likely that "mainstream" education will often come to the professor's attention because the ordinary is seldom a topic of everyday conversation; rather, the new and innovative generate most discussions. Therefore, the impression is conveyed, however harmless the intent, that all is well with the educational world.

In 1968, the Committee for Economic Development (CED) came to a conclusion similar to that of the critics, when it undertook the writing of a national policy statement on innovation in education. The CED has as its purpose the study and formulation of policy on matters of grave national concern. After completing its study, the Committee saw fit to issue as the first (of four) imperatives that "the American School must be better organized for innovation and change."[1] The text which followed the imperatives stated:

The future of the schools depends in large part on whether they can overcome in educational policy and practice what is frequently an extreme conservatism and a strong resistance to change. This depends in turn on whether they can develop a genuine openness to experiment and innovation. This is difficult because the conservatism of the schools has been a natural response to society's expectation that they reflect dominant social opinion and that they perform an essentially conservative function.

The experimental activities of the schools must be designed to protect children and youth from the negative effects of experimental failure, but success will

---

1. Committee for Economic Development, "Four Imperatives for the Schools . . . ," *Innovation in Education: New Directions for the American School* (July 1968), p. 13.

come only if the schools encourage new and revolutionary ideas and are willing to question even the best-established educational traditions. It is evident that the major deficiencies of the American schools cannot be remedied simply by refining the customary organization and procedures of classroom instruction or by extending the work of the conventional schools.

We are convinced that reconstruction of instructional staffs, instructional patterns, and school organization must lie at the heart of any meaningful effort to improve the quality of schooling in this country.[2]

Hardly a clearer or more concise statement of the need for change could have been written.

To be sure, considerable innovation does occur. Within those districts in proximity to colleges of education, within a comparatively few other especially enlightened school districts, and within an occasional isolated school, change does come about. But the sum of these, in comparison to the remaining number of districts and schools, is small indeed. Thus, part of the rationale for innovation is that there is much room for improvement.

Further, and to continue with an earlier point, innovation or change is, in itself, "good." Innovation brings with it a lively, often electric, school atmosphere. Teachers and principals often seem almost to become crusaders for a new idea. Everyone connected with the school becomes visibly affected: Superintendents include the innovative school on the itinerary of important visitors; principals offer tours to anyone who is interested or who will take the time to listen; principals and teachers are flattered to be invited to address teacher and administrator workshops. The excitement is easily transferred to students and often even to parents. "We've really got something going on in our school" seems to frequently be most everyone's attitude.

But notice not a word has been said about whether the particular innovation was any good. The point here is that with normal safeguards the nature of the innovation is totally incidental. Obviously bad ideas, will usually not generate significant support and will almost never be adopted; so within a broad range, the precise nature of the innovation is relatively unimportant.

The supportive evidence for this somewhat bold statement was summarized a few years ago by Richard Foster, now superintendent at Berkeley, California, in a speech before the Association for Supervision and Curriculum Development. Foster accurately summarized the existing research and a massive Rockefeller Foundation study of the comparative advantages of the various curricular and instructional techniques, with roughly the following statement (paraphrased):

2. *Ibid.*, p. 14.

What the researchers have found is that with Method A 50 percent of all students perform above the national median and 50 percent perform below the national median, whereas with Method B, the opposite is true.

The synthesis of the conclusion made from all this research is that if a teacher is convinced that Method A is superior, it will be. If she is committed to a new idea it will work.

Nothing has developed in educational research to disprove Foster's thesis in any form.

Foster's major conclusion is precisely what would be predicted from a segment of the research in social psychology. Probably the most famous of all such research, because it was the landmark study giving rise to the human relations school of management, was the series of experiments conducted at the Hawthorne Plant of Western Electric at the outset of the Great Depression. The reader will recall that the conclusion drawn from those experiments led to what has come to be known as the Hawthorne Effect, which, simply stated, is that people respond favorably to personal attention. At the Hawthorne Plant members of a small wiring room work force were subjected to various treatments, some of which (e.g., increased lighting and time off), it was anticipated, would cause increased productivity, while others (e.g., decreased lighting and time off), it was predicted, would cause decreased productivity. But all manipulations resulted in production gains and the major conclusion drawn was that performances improve when workers observe that they are receiving special treatment. When something new is being tried with them, they feel quite special. Foster's conclusion regarding teacher commitment to a new idea, which she or her group alone have been given the opportunity to implement, is an illustration of the Hawthorne Effect. Teachers respond with great enthusiasm to being singled out for innovation and their enthusiasm carries over to a generally enlivened school atmosphere and success for the innovation.

There is yet another component in the rationale for innovation, a component that is solidly grounded in learning theory. The theoretical principle is that the varying of stimuli enhances learning, or in less abstract terms, individual learning is promoted through variations in instructional techniques and in instructional settings. One gets bored if all instruction is presented via the lecture, the textbook, or for that matter the computer. One also benefits from a change in the environment. Who has not found it helpful to leave the office and go home to work on that report?

Thus we need new ways of doing things in education, new ways of presenting instruction, new ways of organizing the curriculum and new ways of breaking up the day. And we need to vary the educational setting, to move students from module to module, learning center to individual study carrel, gymnasium to laboratory, and so on. We need to use the lecture and the discussion, show films and filmstrips, use the chalkboard and the overhead projector, the computer and the textbook. In short, we need to innovate.

The framework for this chapter is innovation within *a systems approach*. Systems education at its most basic level requires a statement of purpose, manifested ultimately in concrete objectives, and the fulfillment of which can be observed and evaluated. In simpler language, systems education mandates a statement of direction and magnitude. It is based upon the principle that you must know where you want to go if you are to reasonably expect to get there. Robert F. Mager's parable poignantly demonstrates this point in a humorous fashion:

Once upon a time a Sea Horse gathered up his seven pieces of eight and cantered out to find his fortune. Before he had traveled very far he met an Eel, who said,

"Psst. Hey, bud. Where 'ya goin'?"
"I'm going out to find my fortune," replied the Sea Horse, proudly.
"You're in luck," said the Eel. "For four pieces of eight you can have this speedy flipper, and then you'll be able to get there a lot faster."
"Gee, that's swell," said the Sea Horse, and paid the money and put on the flipper and slithered off at twice the speed. Soon he came upon a Sponge, who said,
"Psst. Hey, bud. Where 'ya goin'?"
"I'm going out to find my fortune," replied the Sea Horse.
"You're in luck," said the Sponge. "For a small fee I will let you have this jet-propelled scooter so that you will be able to travel a lot faster."

So the Sea Horse bought the scooter with his remaining money and went zooming thru the sea five times as fast. Soon he came upon a Shark, who said,

"Psst. Hey, bud. Where 'ya goin'?"
"I'm going out to find my fortune," replied the Sea Horse.
"You're in luck. If you'll take this short cut," said the Shark, pointing to his open mouth, "you'll save yourself a lot of time."
"Gee, thanks," said the Sea Horse, and zoomed off into the interior of the Shark, there to be devoured.

The moral of this fable is that if you're not sure where you're going you're liable to end up someplace else.[3]

Objectives or goals are the *sine quo non* of a school's existence. The schools have reasons for being and their goals should reflect those reasons. Policy decisions made by administrators and teachers must be consistent

---

3. Robert F. Mager, *Preparing Instructional Objectives* (Palo Alto, California: Fearon Publishers, 1962).

with goals. This is essential down to the final application stage, which is the writing of the behavioral objectives used to guide every teacher in each days' instruction.

By definition the systems approach demands the viewing of instruction as a macro-cosmic affair involving the joint participation and planning of the total school staff. The teacher ceases to be the dictator of her classroom but instead becomes a respected and vital member of the instructional team. Thus the systems theorists take serious objection to the notion of 30 to 40 students contained in a box and taught by a single teacher; because implicit in such a notion is the discrete, insular nature of each classroom—clearly a contradiction to systems education.

Some would say that the identification and listing of instructional goals in a meaningful form is an impossible task for the schools. This is indeed a difficult problem though a soluble one as demonstrated by the efforts of the Committee for Economic Development which has gone the first and most difficult step in operationalizing instructional goals.[4] From here, the task is one of breaking these goals down into more and more specific and applied objectives.

As a final introductory note a further word about the utilization of educational research is in order. Although it has been pointed out that, considered as a whole, research fails to demonstrate the superiority of any specific instructional method or curricula, this is not intended to imply the rejection of research findings. Quite the contrary. What is desperately needed and what is suggested for education is a more analytic attitude toward empirical methods. Hopefully, the time is passing when educational decisions will be made on a "how do you feel about that program?" basis or on the whimsical reactions of an administrator to a super-salesman. Systems education depends for its effectiveness upon a careful evaluation of results. The philosophy necessary to proper assessment of program results is the same one required in the selection of an innovative approach. Foremost is an open mind that does not reject out of hand the experiences of those hundreds of schools which have found the value of innovative programs. It is the major task of leadership to provide the enlightenment necessary to a program of change.

In way of summary, the Committee for Economic Development provides a series of questions to be used in judging the value of technical resources:

Can the proposed technique be effectively employed in the cultivation of an open, inquiring mind? Or does it tend to produce conformity, dogmatism, and regimentation of thought?

---

4. Committee for Economic Development, "The Goals of Instruction," *Innovation in Education,* p. 33.

Is it capable of communicating and facilitating an understanding of complex concepts? Or is its usefulness limited to the management and manipulation of simple ideas?

Is it capable of cultivating sensitive insight, originality, analytical facility, and creative intellectual skills?

Can it be employed to induce and deepen artistic and moral sensitivity and appreciation?

Do the benefits gained justify the costs incurred? Is the initial cost affordable?[5]

A SURVEY OF INNOVATIONS

This section includes a survey of innovative instructional techniques, supporting technology, and innovations in administration. By no means is a complete listing given, but major innovations and illustrations of general forms are included.

Instructional Television

An instructional television system involves more than the electronic hardware itself. Instructional television (ITV), when properly conceived, includes the television instructors, the technical supporting staff under the control of educational managers, and classroom professional and paraprofessional personnel. When properly designed, the system is *not* a supplement to classroom instruction—instruction to be tuned in or out at the whim of the teacher—but has as its unit of operation the instructional *system* rather than the classroom. Thus, control of the system does not rest with the teacher; it rests within the systems team. The portion of the team that is external to the classroom performs its periodic function of program presentation, as jointly planned with classroom teachers; and then, through personal interaction with students, classroom instructional personnel perform their assigned functions. Research has clearly shown that the tuning in of isolated programs or the periodic sampling of portions of a sequenced course is not effective use of instructional television.

In addition to regular classroom learning, instructional television has been used effectively in reaching students at home. Both the invalid and the worker whose schedule conflicts with normal instructional hours have been primary audiences. ITV has been used for in-service education of industrial workers including professionals in need of up-dating their knowledge and skills. It has been used to present highly specialized curricula and to extend the effectiveness of particularly exemplary teachers.

---

5. Committee for Economic Development, "Judging the Value of the Technical Resources," *Innovation in Education*, p. 45.

But the most effective and efficient use of ITV has been in cases where a unique educational problem has existed, and herein lies the key to selecting the conditions for its application. ITV has been used where chronic shortages of qualified teachers or teacher specialists have existed (especially in certain foreign countries) or where the special conditions have dictated a non-conventional instructional approach. ITV has *not* been particularly successful in the case in which an administrator finds himself the unexpected recipient of some television equipment and merely "wants to get some use out of it," or merely wants to dabble with TV equipment because to do so is fashionable. Under such circumstances ITV has not had any particular impact. No unique situation or problem existed; no mandate was present.

Nor was it likely that other essential conditions existed. Successful use of instructional television requires careful, lengthy planning, which involves teachers and all other prospective team members. It also requires considerable financial and leadership support.

If ITV is to be efficiently used, the economics are such that implementation must be generally widespread. Colombia, for example, provides ITV to 250,000 students at a cost of 5 cents per student viewing hour; in Hagerstown, Maryland with 21,000 pupils the comparative cost is 20 cents; Samoa, with 7,000 students spends a comparative 60 cents; and in a certain African nation having only a few thousand pupils, the cost is $2.00 an hour per pupil.[6]

However, there is more to the economics of using ITV than mere economies of scale. Within the United States, where 10 million school children and 600,000 college students view ITV, geography and decentralized control become the determining factors. The units for cost considerations involving scale become the school district, the separate college, the state system.

Instructional television can be of benefit when available teachers are poorly trained or in short supply, when a special or unusual talent is to be shared, when lesson preparation requires extravagant amounts of time, when materials are too expensive or dangerous for ordinary classroom use, and when great detail is desired via close-up viewing. ITV is less appropriate to do things classroom teachers can do well, when students must perform manual skills, when classroom interaction is to be the basic instructional method and when overall costs are prohibitive.

6. Wilbur Schramm, "Instructional Television Here and Abroad," *The Schools and the Challenge of Innovation* (Committee for Economic Development, 1969), p. 264.

Although presently the economics of ITV are such that little money is likely to be saved through its use except where sufficient economies of scale can be gained, as more subscribers are gained costs will go down. Also, whereas unit costs of ITV equipment continue to drop as mass production techniques become practical, the costs of professional labor continue to rise. At some point the cost curves will intersect, not only in ITV but in most other innovative instructional modes requiring large capital outlays.

Team Teaching

Instruction utilizing the team concept has come about in response to the recognition that teachers have differential talents. Some are more competent in language arts; others are more capable in science, or in mathematics, or in social studies. Some are best at leading and others are best at following. Theoretically team teaching allows for the optimization of each teacher's strengths.

Although proponents of team teaching argue that the method results in a more individualized curriculum and the more flexible use of time, in reality these things do occur but they are more correlates of teaming as it is practiced, than they are necessary outcomes. The school staff which is committed to team teaching almost always subscribes to the entire "package," and because teaming is clearly a systems component, these correlates do occur. But they occur as side effects of the systems approach, not as necessary outcomes of team teaching *per se*. When emphasis is taken off the *individual teacher* as the unit for organization, and concentration is upon the *instructional system*—as is the case in team teaching—then, as would be expected, media cease to be merely ancillary in nature, objectives come to be set and met by the total staff, "clock watching" comes to be a nonsense activity, and instruction comes to be individualized. Non-grading, a rediscovered "innovation" not dwelt upon here, is another natural outgrowth of the team approach to instruction.

Although prior to experience with teaming, teachers are often under the mistaken impression that the technique will afford them more free time, such is not the case in practice. The reason is that any free instructional time plus additional leisure time stolen from before and after classes must be devoted to planning. As in any systems approach, members of the instructional team must meet to set objectives, determine the means for their accomplishment, decide upon evaluation methods, and appraise the results of the evaluation. Scheduling, curriculum development, and grouping are matters for daily deliberations. Communications among team members is not only time-consuming but is a necessity; in fact, the failure of communications is probably the most troublesome and most frequent concern related to the team approach.

Team teaching is totally consistent with, indeed demands, another organizational innovation—differentiated staffing. Teaming requires an authority structure that is modestly hierarchical. Someone must be in charge and para-professional personnel are requisite to the efficient utilization of the professional staff because of the varied, simultaneous activities occurring within each team-teaching pod (a cluster of modules).

The delightful side-effect of differential staffing is the development of an incentive system for teachers. In the view of the author, one of the most negative features of the present career system for teachers is the lack of a meaningful ladder for advancement within the instructional system. At the age of twenty-one, or thereabouts, the beginning teacher is placed upon a salary schedule from which he may not deviate regardless of his excellence. He can, of course, gain additional academic training, but for all practical purposes he is locked in place for life. What self-respecting, ambitious person worthy of himself would tolerate such a system? The answer, of course, is that few do. Most* either change careers or seek upward mobility in education through whatever means are available. However, most open channels for self-betterment are counter-productive to excellence in instruction.

The means by which capable and ambitious teachers seek advancement within education take three forms. Some seek "promotion" to the advanced classes—senior literature rather than freshman grammar; trigonometry rather than general math; honors psychology rather than American history. The effect is to leave the important basic skill courses to the inexperienced teacher. Others prefer "advancement" from the inner city schools, where they are most likely to begin teaching, to the suburbs and to the elite neighborhoods. Here student values will be more closely akin to teacher values, but again the more difficult task will be left to the novices. The final "promotion" is to administration, thus taking the more competent and ambitious out of the direct instructional process altogether. That excellence in administration is needed is, of course, undisputed; however, "robbing Peter to pay Paul" is hardly the answer.

Differentiated staffing vis-à-vis team teaching facilitates a reward system within the instructional systems. Individuals can begin, perhaps immediately upon high school graduation, as teacher aides. Working part time, they can launch into an undergraduate teacher preparation program,

---

* It is fully recognized that these statements do not apply well to most women for whom society has made the classroom one of the few satisfying work environments within their reasonable expectations. Hopefully, times are changing.

emerging a few years later as apprentice teachers. Whatever the labels assigned, most can eventually advance to regular teacher and finally, for some, to master teacher status. In this way, good teachers can be retained and good teaching rewarded.

Team teaching is probably the recent innovation having the greatest acceptance and application in the field. An estimated 41 per cent of high schools had developed team teaching in some form by 1967.[7] The percentages of elementary schools involved is less, but where existent, full adoption is much more likely to have occurred. The compatability of team teaching with other elements consistent with the systems approach—programmed instruction, individualized instruction, differential staffing, systems architecture and construction, flexible scheduling, etc.—promise the continued growth of this promising innovation.

Programmed Instruction

Programmed instruction is a predesigned, step-by-step, logically arranged, self-tutoring, instructional technique. The concept of programmed instruction focuses upon the learning process and is based upon behavioral sciences learning research. Programmed instruction is an attempt to capitalize on this research through "teacher proofing" the instructional process. Notwithstanding the rhetoric of salesmen, when properly utilized, control of student learning is taken from teachers and placed in the hands of scientists in the form of the programmed text.

Programmed instruction has most often been used in elementary and secondary schools to present total courses, teach portions of courses (units), and for remedial and enrichment purposes. Mathematics and sciences are the major subjects which have been programmed although programs in other subject areas are beginning to grow in number. Available data reveal that in 1963, 36.3 per cent of those schools surveyed used programmed instruction, but that the extent of use within schools was very limited.[8] Although in 1966, elementary school programmed materials purchased represented less than one per cent of textbook sales, this portion of the market is expected to grow to 10 per cent by 1976.[9]

Programmed instruction lists as its great advantage, the individualization of instruction. In fact one approach—individually prescribed instruc-

---

7. James A. Meyer, "Teaming a First Step for Interdisciplinary Teaching," *The Clearing House* (March 1969), p. 407.

8. Lincoln Hanson and P. Kenneth Komoski, "School Use of Programmed Instruction," in *Teaching Machines and Programmed Learning II: Data and Directions*, Robert Glaser, editor (National Education Association, Washington, D.C., 1965), pp. 647-684.

9. Robert Glaser, "The Design and Programming of Instruction," *Challenge of Innovation*, p. 208.

tion (IPI)—is designed purely for this purpose. In IPI, as in other forms of programmed instruction, students are allowed to proceed at their own rate (money savings though time savings is another claim) and the program can differentially respond on the basis of student inputs. The flexibility with which programs can react, however, is limited unless combined with a computer system. Non-computerized programs cannot possess the vast array of student background data available to teachers. Thus, it is generally acknowledged that teachers are potentially more responsive to individual differences than are programmed texts.

The entire question of individualization in the schools, however, is really a paper issue, with programmed instruction probably at least as sensitive and responsive to individual differences as are teachers, although the latter do indeed have a greater potential in this regard than does *non-computerized* programmed instruction. But teachers do not really respond in significant part to individual differences because the system militates against it. The system ostensibly provides for individual differences by such practices as tracking, ability grouping, student repetition of courses or grade levels, and remedial and enrichment courses. This is, however, only the grossest kind of differentiation and probably does more damage than good by creating the false impression that individual differences have been met and that no other action for individualization is required. The traditional instruction can then be justified, with the delusion that each student is receiving a tailored education. Teachers generally share in and further this educational hoax by failing to take proper instructional action partially on the presumption that the organization of the curriculum has met individual needs. However, it must be recognized that it is simply not realistic to assume that teachers can truly individualize instruction without electronic assistance. The daily clerical tasks of critiquing separate and distinctly different student assignments, making differential assignments, and keeping track of each student's place in his work would consume almost all class time, leaving little room for personal assistance.

There are several difficulties with the use of programmed instruction. To teachers, the method seems highly technical and impersonal; it is seemingly the work of engineers more interested in efficiency than personalism. Teachers, unfamiliar with the role demanded by programmed instruction, often attempt to keep groups of students progressing at similar rates so that intervention in the instructional process is convenient. They attempt to find new ways of serving as subject-matter mediators, such as forming and leading discussion and review sections. Teachers are accustomed to being the center of attraction and in a programmed setting their sense of self-worth can suffer initially. Finally, they do not enjoy the immense bookkeeping chores that are required.

A second difficulty in the use of programmed instruction is in the quantity and quality of materials available. Few materials exist that are capable of meeting classroom needs although this picture is changing.

Another difficulty involves the role of administrators who are often perplexed as to their supervisory function in a programmed instruction setting. Richard O. Carlson's research into the effects on administrators of short-term programmed instruction reveals that principals do not know what to look for in supervising teachers engaged in this instructional process. As a result, they abandon the observation of teachers and come to rely upon feedback from students. Their efforts at observing instruction generally consists of noting whether the teacher circulates around the room and whether she is orderly in storing and maintaining supplies.[10]

Computer Assisted Instruction

CAI, as computer assisted instruction is more popularly known, is a sophisticated outgrowth of programmed instruction. CAI obviously involves the computer; programmed instruction is ordinarily conceived to imply no particular hardware. Programmed instruction is the teaching mode basic to a computer assisted instructional system; that is, CAI utilizes the concepts and principles of programmed instruction. But it also does more. CAI adapts to the learner via two-way communications. It can adjust to the pattern of learner responses on any given sequence of materials. Whereas programmed instruction is designed to present a particular subject through a series of simple-to-complex "courses," (e.g., 9th grade algebra, "higher" algebra, college algebra), CAI can present a subject (i.e., all of algebra) in a single unit, varying the form and sequence in accordance with pupil input data.

Because of these qualities, CAI is the single instructional technique or device that can truly individualize instruction. Programmed instruction cannot respond differentially except in the simplest ways. Most other media are purely for group presentations. (CAI can present multi-media forms on a screen in accordance with individual needs.) Even the teacher, who if she is unusually alert and capable may have the *potential* for responding to students on the basis of their personal characteristics and daily progress, most likely will lack either the inclination or the time to provide truly individualized instruction. CAI falls short of the mark at present only because the state of the art is relatively primitive, but it has both the potential and the means for fully meeting this instructional challenge.

10. Richard O. Carlson, "Programmed Instruction: Some Unanticipated Consequences," in Glaser et al., *Studies of the Use of Programmed Instruction in the Classroom* (Learning Research and Development Center, Pittsburgh, Pa., 1966), pp. 121-131.

Indeed, here is the explanation for the somewhat mild endorsements afforded CAI today; the present state of the art is presumed to be the maximum potential. In actuality, CAI is at a stage roughly comparable to the discovery of gunpowder in modern weaponry. Few programs have been written; few skilled professionals and technicians are available; little individualization has occurred; and costs are still high although as in the case of instructional television, economics of scale and rising labor costs are rapidly changing the comparative cost picture. Presently CAI costs in lower schools are approximately four times conventional system costs and are roughly equal to conventional instructional costs in higher education.[11] Patrick Suppes estimates that per terminal costs, if connected to a computer in the district office, would be about $1,000.[12]

In addition to its great capabilities for individualizing instruction, CAI has unusual potential for evaluating student performances. Student responses to intermittent questions are stored and can be retrieved for assessing student progress. These data are also valuable in analyzing all components of CAI courses and lessons down to the single sentence or clause in the presentation. In this way not only may presentations be upgraded, but learning processes may be more readily understood and teaching methods, techniques, and sequences may be evaluated. The boon of this CAI feature to research in learning and instruction has barely been tapped.

Even though many students can work on separate consoles simultaneously, CAI gives the learner the feeling that the instructor (the computer) is responding to him on a one-to-one basis. The computer is not recalcitrant, is not condescending, and does not respond with a pained expression if an answer by a slow learner is incorrect. On the contrary, the CAI program can be genuinely supportive of gains made by even the slowest of learners and if the research showing the transfer of learner success in one area to success in other endeavors is correct, truly significant gains can be made through CAI.

Current uses of CAI include almost all conceivable subject areas but are focused in mathematics, sciences and language arts, with some expansion into the social sciences and humanities. Programs are typically of unit length although a few dull courses are available. Remedial and inservice populations seem to be favorite targets for which materials are being designed.

11. Lawrence M. Stolurow, "Computer-Assisted Instruction," *The Schools and the Challenge of Innovation* (Committee for Economic Development, 1969), pp. 308-309.

12. Patrick Suppes, "The Computer and Excellence," *Saturday Review*, 14 January 1967, p. 50.

The "New Curricula"

Beginning in the early 1950's with the "new math," many subjects of the school curriculum have been revised along, what is for the most part, new and drastically altered lines. Sometimes there have been not one, but two or more new curricular approaches to the teaching of each subject, e.g., chem study and the chem bond approach. Following elementary school mathematics, curricular reforms in the sciences appeared. Curriculum projects, staffed by a number of the nation's most eminent university scientific personnel, developed in physics, chemistry, biology, earth science, and physical sciences *in toto*. Before long revisions extended down into the elementary schools, with a feature of these curriculum projects becoming a K-12 format. In the meantime developments took place in English, modern foreign languages, geography, anthropology, economics and other social sciences. An educational revolution had truly occurred although in relative terms the portion of classrooms affected remained small.

The new curricula come complete with new texts, study guides, achievement tests, films, filmstrips, transparencies, and individual student equipment and materials. Everything needed has been provided by enterprising publishers and supply houses. The teacher can follow the presentation sequence of handy teacher guides almost word-for-word, departing only where she pleases, if at all. Indeed, these courses have many of the appearances of programmed materials.

The new curricula often seek to gain student attention by beginning with an historical case study of an interesting event or of the life of some significant individual before proceeding to the teaching of conceptual understandings usually using an inductive or inquiry approach. Emphasis is removed from the traditional learning of facts; understandings, generalizations and problem-solving are the main objectives. Textbook information is presented in carefully thought-out sequences and is supplemented by a smattering of eye-catching drawings, diagrams and photographs. Sections are climaxed by thought-provoking propositions and questions requiring basic understandings. Inquiry-oriented discussions, experiments and individual research projects supplement the learning process throughout the total course.

All-in-all the new curricula are impressive in their aim and approach. They are totally consistent with the systems approach to education, being longitudinal in nature (often K-12); being based upon clearly defined objectives; and being completely self-contained, yet complete. If the teacher can be made to resist the temptation for intervention and excessive mediation, the new curricula provide an excellent wedge into a systems approach and into the institutionalization of innovation—a necessary process for excellence in modern school administration.

Of course, as with any new approach, there are problems to be resolved. Perhaps the most serious of these in the case of the new curricula is the challenge of overcoming the old traditions. For generations students have sat placidly back, memorized facts, and beaten the system by performing well on recall examinations. But the new curricula demand higher level thinking on the part of students and organic examinations reflect this objective. One result of the old traditions and of what is probably an inherent problem of the emphases of the new curricula has been that whereas average and upper ability students seem to achieve quite well under the new curricula, below average students often do not. Apparently, the transition from recall to understandings is troublesome for these students who perform relatively less well, in comparison to the better students, than they did in traditional courses. It would seem that the variance in achievement among students is magnified when higher level thought process are the criterion. What this means, however, is probably not that the new curricula are less suited to below average ability students than were the traditional curricula, but simply that the objectives of each approach differ.

Supporting Technology and Administrative Innovation

In the 1960's federal legislation made possible the vast purchase of all kinds of technological, or as they are more commonly known, audio-visual aids. In a matter of only three years, expenditures more than doubled.[13] Schools which formerly contained no more than one or possibly two 16mm projectors and an overhead projector often found themselves in possession of several additional film projectors, an overhead projector in almost every room, opaque projectors, television cameras and receivers, video taping devices, bioscopes, tape recorders, language laboratories, film loop projectors, record players, film strip projectors, and electronically controlled individual study carrels. To be sure, not every school participated in the federal program to this extent but many did and billions were spent in this manner.

But what of use? Indications are that utilization of media devices remains very low, and herein lies a problem worthy of the best administrative techniques. Teachers apparently fail to use most media devices for a number of reasons, the foremost of which is apparatus inflexibility. Most

---

13. Thomas W. Hope, "Nontheatrical Films," *Journal of the Society of Motion Picture and Television Engineers,* Vol. 73-76 (1964-1966).

media are designed to do only one kind of thing and almost all require considerable planning. A teacher can wheel in a 16mm projector and show a film, but that is essentially all she can do with it. Furthermore, she must select the film from a catalog and place an order; she must schedule a projector and screen and probably find an operator. Similarly, television programs must be viewed at a particular time; and the remainder of the day must be scheduled accordingly. Contrast this with the flexibility of the most basic audio-visual aid of all, the chalkboard. Philip W. Jackson provides the narration:

The blackboard is literally at the teacher's fingertips. He can write on it, draw on it, immediately erase what he has written, or preserve it for days. He can scrawl key words on it, produce a detailed diagram, or write out a series of essay questions. He can use the board himself, or ask his students to use it. He can place material on it in advance, or use it to capture the fleeting and ephemeral thoughts emerging from a discussion. Given this flexibility, it is no wonder that the chalk-smudged sleeve has become the trademark of the teacher.[14]

The message for the administrator is that the traditional school organization is simply not readily compatible with technological innovation. Media have traditionally been under the control of the teacher and have been used in group instruction. In the systems approach media are used primarily on an individualized basis and are an integral part of the total educational system. They are not for supplementary purposes, but are central and thus demanded by the system. The use of computer assisted instruction, programmed instruction, and system-integrated instructional television illustrates this point.

The systems approach, with its specified learning objectives, suggest another principle for the use of technological devices. Although overzealous salesmen have represented such devices as broadly superior regardless of the learning objective sought, research indicates that for optimum learning the form of the learning sought—e.g., visual, auditory—should correspond with the form of presentation.[15] In other words, if visual identification of objects is desired, then objects should be presented in picture or concrete form. Few media forms—CAI being an exception—have multi-sensory capabilities; the implication is that a particular device should be used to fulfill a specific need. This does not mean, however,

---

14. Philip W. Jackson, "Technology and the Teacher," *The Schools and the Challenge of Innovation* (Committee for Economic Development, 1969), p. 129.

15. C. R. Carpenter, *AV Communication Review,* Vol. 1, 2, 11 and 14 (1953, 1954, 1963 and 1966).

that the device is only a supplement; it is an integral part of the instructional system.

In addition to technological devices, innovative instructional techniques have been accompanied by exciting new ideas in architecture. The unifying principle has been systems construction, which is aimed at providing the flexibility in facilities required by systems education. Thus the move is away from the "egg-crate" classroom, which has seen students playing a game of musical chairs with the hallway bell corresponding to the cessation of the music in that old parlor game. Systems architecture renders obsolete dark halls and standardized rooms, equipment and materials. Thirty-five ceases to be the magic number of pupils required to compose an instructional group.

In the place of traditional uniformity is the architectural principle of flexibility. Buildings are constructed in such ways as to allow varying class sizes, special learning centers, individual carrels and multi-purpose rooms. Construction patterns utilize pre-engineered modular components, consisting of long clear spans, lacking interior supports and including the capability for easy installation of heating-cooling and lighting systems. The concept is long-term adaptation to change so that no longer will a temporary educational fad or the current style of instruction dictate future configurations.

Arrangements of modules presently in vogue do indeed reflect the varying instructional systems. The classrooms-without-walls concept has accompanied team teaching and flexible scheduling. In this design, modules are usually arranged in open clusters of three or fours to facilitate the easy flow of students from module to module, where the various members of the team group and re-group the students for quasi-individualized instruction. Most clusters include a large adjacent open space, where students can spend time at the multitude of different individual and group stations.

Flexible scheduling is an administrative device for responding to the flexibility demanded by the systems approach. In this process classes are of varying duration and composition because the individual student rather than rooms or teachers is the unit for scheduling. A given student might attend a Monday and Wednesday morning 40 minute chemistry lecture, devote all Thursday afternoon to chemistry lab, spend the remainder of Monday morning in English literature discussion sessions and an honors class in sociology, devote Friday afternoon and various other shorter time spans to self-directed study, and so forth. The point is that class time spans vary with functions to be performed as do the facilities within which instruction is received. Unfortunately, under present systems, schedules are usually set for the year or term and cannot be easily altered on a daily basis,* which is the essential condition for true flexibility. Another caveat is that flexible scheduling by no means guarantees teachers will do anything different than they did under the old system. This is, of course, where administrative leadership enters in.

The linear school which largely came into being in response to rapidly rising urban land costs, is another architectural arrangement compatible with the systems approach. In this design classrooms may be blended in with commercial facilities, community agencies, and playground facilities; often they are built over or under transportation arteries or in conjunction with private business establishments.

Yet another facilities innovation is the educational park, which combines on one site a number of elementary schools, usually a fewer number of junior high schools or middle schools, and a single high school. The perceived advantages are multiple use of resources (e.g., swimming pools, playgrounds, libraries) and the possibility for the reduction of *de facto* segregation.

Performance Contracting

The entry of private enterprise into elementary and secondary education is neither an instructional nor a support innovation; yet it is a relatively recent development that may turn out to have enormous implications for both teaching and administration. Performance contracting and allied administrative devices involve an agreement between school districts and private entrepreneurs that the latter will perform some educational service and be paid for it. Often, if the service is not fully provided, that is, if the students fail to perform as specified in the agreement, the contractor will pay a penalty and will collect only for students who meet the designated standard. The essential elements of a typical program are:

1. Students who are below standard in basic skills or other priority areas are to receive a training program for an agreed-upon period of time in a portion of the school plant. The students remain in the total school program to receive other school benefits.

2. The contractor agrees to train school personnel so that the school system can carry on the successful practice after the project is terminated.

---

* Terminology is important here. Modular scheduling, variable scheduling and modular flexible scheduling are often used interchangeably in the literature; but to the sophisticate they mean different things. Considering the normative case the statements made above are accurate; however, in a number of schools greater daily flexibility exists in the form of the varying uses of student free time. In a few other cases schedules are rewritten periodically throughout the term.

3. The contractor agrees to be paid only on the basis of a stipulated amount of money for each student who successfully completes the training program.

4. A penalty is assessed for those students who do not achieve specified performance levels.

5. A stated time after the termination of the project, school officials have a right to reassess student performance. If it is less than the specified level achieved, a penalty may be assessed.

6. The school system, not the contractor, selects the students.

7. The training program of the successful bidder must be cost-effective and not labor-intensive.

8. An independent audit is mandated as the basis of payment.

9. The contract stipulates a "turnkey" or system-incorporation feature to insure general use of a successful approach so that the program continues after the contractor has terminated his services.[16]

The key concept of performance contracting is accountability in a systems approach to education. Accountability has come to be demanded by a public that is no longer willing to accept education on faith. The public is demanding results for their ever-increasing tax dollars, and astute administrators see performance contracting as a means of meeting the public mandate. In performance contracting, clearly-defined behavioral objectives are established in keeping with the systems approach and contractors are held accountable for results.

Contracts awarded have most frequently been for only a certain, small portion of the school curriculum. Usually, a single academic area requiring the attainment of very basic skills has been the subject of the agreement; the development of second-grade reading skills, or of sixth-grade arithmetic skills are examples. The focus has typically been upon remedial work or the somewhat difficult task of skill attainment by disadvantaged students. Occasionally, an entire school or district (e.g., Texarkana and Gary, Indiana) has let contracts in skill areas at all grade levels or in all subjects in selected schools.

The development of performance contracting has proceeded fairly rapidly. From an initial expenditure of $250,000, performance contracting has expanded to a 100 million dollar enterprise, encompassing hundreds of school districts, although generally the saturation within districts has been modest. Recently, the United States Office of Education has indicated less performance contracting for the future.

16. Leon M. Lessinger, "Accountability in Education," *Resources for Urban Schools: Better Use and Balance* (Committee for Economic Development, 1971), pp. 27-28.

Expansion from the basic skill courses has not been rapid; to the present, contracts have been awarded for the teaching of those subjects having goals that can be clearly stated in concrete terms, e.g., "reading up to grade level." "Developing better citizens," or "bringing about student social development" are objectives which for the most part private enterprise has not yet been willing to tackle.

THE ADMINISTRATOR'S ROLE IN INNOVATION

The central task of administration in any system is leadership. When innovation is desired, the demand for leadership is absolute; there will be no change without it. The administrator, who merely chides his staff for their lack of an innovative spirit or exhorts them to be more change oriented, is doomed to fail. Change is the most difficult of all conditions to bring about. It requires hard work, sustained efforts and perseverance. It requires a leader who is not already overburdened with day-to-day routines, someone who is not deeply engrossed in maintaining the organization in its present form. In short, innovation requires a leader.

Any administrator can fill out his day with routine tasks. Parkinson's Law tells us that. However, the educational *leader* delegates the authority for conducting routine affairs and turns his attention to those matters that will establish the tenor and direction of his institution—what the school or school district will "become," what it will seek to accomplish, what distinctive character it will assume. These are the leadership tasks of administration and they cannot be delegated. Subordinates can have major authority in monitoring the halls, making announcements over the intercom, disciplining students, answering to parents, assigning lockers, and ordering foodstuffs for the lunch program. They cannot be given authority for developing institutional values and seeing to the internalization of those values by members of the school staff. Such is the task of leadership.

The Conditions for Innovation

The rationale for this chapter is based upon the human relations theory of administrative leadership. Teachers, second and third echelon administrators, counselors, para-professionals and clerical staff—all members of the instructional team—deserve and expect considerate treatment. Effective leadership mobilizes the instructional staff for the achievement of objectives by enlisting team cooperation. Authoritarian administration must cope with staff resistance, however covert, at every turn.

The educational leader begins his task with the knowledge that the major forces shaping education are history and tradition. "We have always done it this way here" is the phrase perhaps most often heard by the administrator new to an organization. The unconscionable assumption is that things have been done a certain way because there is no better way; otherwise, the better way would have been implemented. As feeble as the merits of this viewpoint may be, it is nonetheless constantly with us. Seldom do we reflect upon the way we do things and the continuation of long-standing procedures *reinforces* the unconscious presumption that our way is the best way. However repugnant we may think we find it, "Don't rock the boat" is a cliché having deep roots and almost unanimous support in many organizations. Most of us merely fail to recognize the phrase in its manifest forms.

There are numerous examples of the seemingly insurmountable traditions of the schools, but traditions appearing to be without educational merit in a changing world. There is the teacher-centered, self-contained classroom; the lecture method; the department structure; the textbook and the workbook; the rigid class schedule; the Carnegie unit; and the standard architectural design, to name a few. There seems to be hardly any important characteristic of the schools in which history and tradition do not play major, if not total, roles. Yet, as seen in the previous section regarding contemporary innovations, there are ample and sometimes overwhelming reasons for calling these practices and traditions into question. Again, this is the task of leadership.

How then does the educational leader proceed? The first step is gaining the full realization of the importance of history and tradition. When the total impact of these forces on resistance to change is clear, the leader will not underestimate the magnitude of his task. When the leader fully comprehends the implications of history and tradition he will understand resistant staff behaviors. Then the more difficult task of making the staff aware of the causes of their behavior begins. The mode for this cannot be authoritarian, such as a sermon given at a faculty meeting where the administrator is clearly in charge, but must be non-directive in manner. Group dynamics is one approach which suggests ways for accomplishing this in an open, non-threatening way; but whatever the technique, staff awareness that their educational practices are largely grounded on little more than history and tradition is essential to a receptive attitude toward change.

The generalized concept illustrated by this specific issue is organizational readiness. The administrator must develop a number of conditions within his organization (the school) in preparation for the implementation of an innovation; otherwise, a new approach will be similar to the proverbial sowing of seeds upon rocks.

Certain aspects of the readiness required may be visible. A new building designed along the systems line, the installation of a computer, the issuance of a flexible schedule—each of these signifies that some new approach is in the offing. And it is amazing how often such developments are installed or announced in the news media before any planning with staff takes place.

Post hoc planning with staff members is easily perceived by them to be little more than a finger exercise—an attempt to legitimize the decisions already made by administrators. It is hardly surprising in such cases, that teachers do not respond wholeheartedly, let alone, enthusiastically, to announcements that they will be "fully involved" in the planning of new programs. The foregone conclusion that a new program will be adopted settles the most important question to be considered. The leadership principle involved implies the inclusion of all staff at the earliest stages of planning if support and commitment are to be expected.

But more than staff commitment is needed; technical readiness is also required. How many times have administrators made the obvious error of buying new equipment and materials and then simply having it dumped on the school doorsteps a few days before September 1? Too many times it would seem. Today, most reputable corporations provide technical support to the purchasers of their products. Many offer teacher preparation and follow-up services at little or no cost to the district. As a minimum, teachers can receive summer workshop instruction and in-service follow-up seminars concurrent with the implementation of the innovation. Where such support is not commensurate with the purchase, the district can and must provide the necessary instruction.

Beyond organizational readiness, the atmosphere essential to innovation and change is one of freedom and acceptance. Teachers must be treated as professionals if they are to be expected to act as such. If the administrator desires more than begrudging acceptance or resignation by the staff to a new approach, he must openly solicit their inputs. If he desires continued success for a new program after the developmental phase and after the program is totally in the hands of the instructional staff, he will have to rely upon their professional commitment. All of this requires an open atmosphere in which staff members can ask questions and receive answers, challenge and be challenged back; in short, the attitude must exist that any innovation will be a joint venture to which all are in at least tacit agreement.

Freedom is the greatest and the conditions are most conducive to innovation when there is opportunity for a fresh start. There is never greater potential for change than on the occasion of the establishment of a new school or the influx of a large group of new teachers. Under these conditions history and tradition are put aside for a time and momentarily at least there is considerable opportunity for innovation. It is a fleeting chance and if not taken immediately it rapidly disappears as the traditions of other paces and other times quickly resume their powerful control. But a new school has no compelling traditions nor do new teachers, and the administrative leader who has an innovative plan can capitalize on the situation. Resistance to a new idea will be at its all time low; indeed, teachers transferred to a new school bring with them an aura of anticipation—the hope that something new and exciting will take place, that the dull routines of the past will give way to new and stimulating activities. If this opportunity is denied, staff morale is likely to be considerably depressed and any eventual attempt at change may be harshly rejected.

Another important condition for inovation is found in the basic receptivity to change by staff members. Hopefully, change is *possible* regardless of faculty attitudes but as any administrator knows, a receptive staff makes the task much less difficult. The obstinate, ultra-conservative, veteran teacher, who actively opposes anything new, is legend around almost any school. Although the expert educational leader can overcome most forms of opposition, the task is much easier if faculty attitudes are positive.

Implied here is the extreme importance of staff selection. Teachers and other staff members are the critical program elements. Their attitudes and skills are the elements that will cause the innovative program to prosper or fail. What principal has not said aloud, "If I only had four more like Mrs. X"? "Adequate" teachers are everywhere and they go through their professional lives doing their jobs in non-descript fashion, making little real difference, as though they never really existed. They fill a position, but over time they rarely if ever have any particular impact upon the lives of students.

For the most part, the habit of school administrators in staff selection has been to fill the position no matter what. Find a warm body at all costs rather than leave a classroom empty, has been the overriding value. And perhaps this position was justified on some grounds with the perennial teacher shortages of the past, but in this time of teacher surpluses, a teacher hired in desperation may well be present and inflicting her damage for thirty or forty years. Multiply this by thirty or forty children per year and the full toll of victims can be estimated. One may argue that if a teacher is truly incompetent she will be terminated, but this is a denial of reality. Teachers are fired only rarely and when they are they are usually employed elsewhere. More often we simply tolerate the incompetent person or if the principal is a man of unusually strong convictions the incompetent teacher may be transferred to another school within the district, there to bring suffering to a new group of children.

That the habit of employing the marginal teacher is ever justified, is open to argument; however, in these times of oversupply there can be no excuse for this practice. With applicant to vacancy ratios running as high as 100:1 in "attractive" districts and at least several applicants for each position in "undesirable" districts (contrary to myths that these districts still cannot attract teachers) there is need to select only the excellent teacher. There is no evidence, however, that this is the practice. Apparently, personnel officers all too often by-pass the change-oriented, enthusiastic, well-qualified teacher for the individual who will be easily socialized to traditional school practices. It is as though school administrators, so accustomed to the "bear market" of the past, do not know how to respond to present favorable teacher supplies.

The message to the change-oriented educational leader is simply to use proper care and caution in staff selection. When change is to be the motif, the "ideal" teacher is not the reserved, deferent personality but is the highly motivated and enthusiastic professional. If the ideal teacher is not available in March, most likely she will be in April or May.

Strategies in Bringing About Innovation

An atmosphere of freedom is also essential to the occurrence of what is probably the most typical kind of innovation—experimentation with a new approach by the individual teacher. The autonomy of each professional to try new things is perhaps the most important condition to be nurtured by educational leaders desirous of change. Indeed a few change agents within a school may be all that is needed for widespread innovation.

Through the use of selective reinforcement, the administrator can hold up the innovator as a model. The innovator can be asked to make presentations at faculty meetings, hold demonstration sessions for other teachers, and keep her door open to others who may be interested in her technique. More visibly, she can be recommended for citation by the school board, suggested for interview by local news media, or recommended for local, regional or national awards. Or most simply, the administrator can merely show his interest and support by spending considerable time in the teacher's classroom, observing, asking questions, looking interested, and offering encouragement. If the motive is not too obvious, other teachers will soon begin to gain the innovative spirit and when they make that very tenuous offer to try something new, the administrator can be ready and waiting to provide moral and resource support.

Dwight Allen tells of how this technique was used in a public high near Stanford.* It seemed that Dr. Allen and his colleagues were seeking means to encourage teacher self-evaluation through video taping and replay viewing. The notion was that if teachers could observe themselves teaching, critical self-analysis and improvement in teaching behavior

---

* From a private conversation with Dr. Allen.

would follow. The difficulty was that teachers were steadfastly opposed to the intrusion of the cameras into their classrooms.

Finally, a single brave soul was found only after Allen's crew promised that no one would be present for the taping. The camera and equipment would be set up prior to class and the cameraman would depart after having secured the equipment in place.

After the recording was made, the teacher viewed her teaching in private. Fortunately for Allen's purposes, the teacher was astounded by her performance—as most persons are when they see themselves on tape or television for the first time. Like most, she was terribly self-critical and immediately took steps toward self-improvement. She repeatedly requested use of the recording equipment for personal surveillance, finally inviting Stanford personnel to assist her in the process of professional improvement.

When this occurred Allen's staff asked the teacher if she would volunteer to relate her experiences at a faculty meeting. By this time the teacher's confidence was so improved she suggested the showing of "before and after" tapes of her teaching. When the "before" tape was shown, the teacher's narration was interspersed with her own laughter. The improvement revealed by the "after" tape was obvious to all.

The message to other teachers was clear and the proof had been seen. The video-taping techniques was helpful in self-analysis and improvement. Immediately, a number of other teachers asked for "private" recordings and viewings; a few confident individuals asked for supervisory suggestions. Within only a few weeks almost every teacher in the school had requested taping of her classes. The eventual total contribution to quality in the school's instruction must have been immense.

A related innovation-inspiring technique is the selection and encouragement of the individual who has a change-oriented profile. The observant administrator will have little difficulty in identifying this person and often a very modest resource expenditure may be the only catalyst required for innovation. Assuming the tolerance of the reader, a personal example will illustrate.

My first teaching position was in a large southern California high school where I taught five sections of biology (although I was a chemistry major). Lacking any special knowledge of the pedagogy of biology teaching, I established a very traditional biology program, using the only approach to biology I knew—the conventional taxonomical method under which I had been taught in high school and college. Had I been teaching chemistry, perhaps I would have implemented a CHEM-study approach —one of the "new curricula" with which I was familiar—but in biology I had no such knowledge.

The school principal very wisely allowed me time to get adjusted, but recognizing my affinity for change (I developed a few classroom innovations new to the school), soon after spring vacation and in an informal discussion he mentioned a new biology curriculum (Biological Sciences Curriculum Study—BSCS) being implemented at a high school in Pasadena. Later, he asked if I would be interested in trying the BSCS approach the following year presuming that he could raise some money for books and laboratory apparatus. In a fashion predictable from the introductory discussion at the beginning of this chapter, I promptly responded in the negative. I had just learned one system, thought it to be pretty good and besides, change was threatening. My Biology Department colleagues, who were equally ignorant about BSCS, reinforced my views; indeed, they warned me that upon my shoulders rested the integrity and autonomy of the department against "non-expert" outsiders. Urged onward by these veterans, I was having no part of this threat to the finest traditions of biology to which I was a recent convert. (And there is no believer like a convert, so they say.)

But the principal knew my weakness and he had one ace up his sleeve. It was spring in California, the school year was at that tedious time, and I had a strong ego. He appealed to all three. Wouldn't I, as "the most enthusiastic and innovative member of the department," like to take a couple of days off—with pay—for the purpose of an informal visit to Pasadena? No obligations. All I had to do was report back on the nature of the Pasadena BSCS program.

The hooks were in. BSCS was observed to be new and exciting; the students in Pasadena were obviously "turned on;" and best of all, there was a chemistry-based version of BSCS. Although that principal never personally reaped the benefits of his efforts because I transferred that summer to a school in the mountains of northern California, I am sure he would be satisfied to know that I introduced BSCS to hundreds of students in the following years.

Here was an example of how a wise administrator could identify an innovation-prone teacher and capitalize accordingly. A free trip to Pasadena and two days off were all that was required to bring about change: total cost, less than 50 dollars.

Another stimulus to innovation may exist in the presence of a near-by school of education. Although some administrators resist any intrusion by teacher education programs into their schools and others merely tolerate from time-to-time a few student teachers, the farsighted leader recognizes the tremendous potential of having college personnel working within the schools. Unlike public school teachers, professors of education often have sufficient time to keep up with new techniques; they have time to read the literature; they usually know how to interpret research findings; and they attend national and regional professional meetings. Further, their reward system generally dictates that they experiment and develop new techniques. Not only does this create people who are change-oriented (although obviously not all are), but also people who will find it necessary to work with the schools in mutually beneficial ways in order to enlist the support needed for the implementation of experiments and new programs. The schools are their testing grounds and collegiate innovators can often infuse the innovative spirit into a total school staff.

In conducting their research, professors of education must often work very closely with a single teacher or group of teachers. These persons often come to consider themselves as members of a research team and as such, they gain commitment to the new idea, program, or approach. In developmental stages they are the logical persons to oversee full field testing in the classrooms. In the full implementation stage they are likely to serve as the demonstration teachers. In each of these roles they can serve as models for change, as the seeds for wide scale innovation.

There are other important benefits of collaboration with schools of education. The presence of outsiders, especially experts, tends to make persons more self-conscious about the way in which they conduct their affairs, about the way they teach, the way they treat students and the way they interact with their colleagues. The viewing of classes by professors or even student teachers or student observers is generally sufficient to shake most teachers out of their doldrums. Teachers who are about to be observed often make careful preparations; they do not want to look foolish before their quasi-peers.

Further, teachers can learn about new approaches by watching professors and can even benefit from observing student teachers. Professors, demonstrating a method or technique for students, also provide an educational experience for teachers. In many cases student teachers probably teach more than they learn by providing teachers with the opportunity to observe the application of the current state of the teaching art.

Another external group from which much can be learned and the seeds of change introduced, is the business community. Those companies which have ventured into the educational marketplace are always anxious to demonstrate a new product for the purview of teachers and administrators. The entry of business into education—even where business has contracted for special services or total school operations—should be welcomed by administrators as a stimulus to innovation and instructional improvement. Business re-introduces the incentive system into education; education professionals are likely to become responsive to change designed to met the competition. Indeed, a principle of change strategy is that crisis and conflict make change much more likely than is possible under "business as usual" conditions.

A final strategy, having perhaps limited application, involves the unusual opportunity for innovation when written materials are in short supply, either due to budget stringencies or supply difficulties, although some administrators have been known to help the process along by failing to place orders on time or by pleading poverty to teachers. The strategy suggested is based upon the realization that teachers often use textbooks and workbooks as crutches and that the presentation of a short lecture with verbal questions followed by a reading assignment and written questions at the end of the chapter or completion of workbook pages remains as the typical instructional format in the American schools. However, the absence of written materials *forces* teachers to innovate and in realization of this a few terribly persuasive or authoritarian administrators have taken away the crutches. Much more of this should be done, especially in the cases of those few blantly lazy teachers.

Facilitating the Innovator and the Innovation

Once an innovation is underway the battle is half over, but it is by no means completely won. The leader's actions continue to be all-important. In this section two kinds of administrative action will be examined: What the leader should do of a psychologically facilitating nature and what he should do in the way of providing logistical support.

The effective leader sets the example. He knows that he cannot advocate freedom and openness in the classroom while administering a "closed" school. He knows that his every action will be appraised for inconsistencies with his verbalized philosophy. Teachers will discount as a bigot the administrator who demands of others what he will not require of himself. The administrator has many opportunities to demonstrate his good faith and the value of his convictions. He can, for example, conduct staff meetings in a democratic way, allowing others to state their opinions, and later carrying out the dictates of the group consensus. The contrast with typical staff meetings, which are used primarily to disseminate to teachers the policies passed by the administration and board, will be so startling as to clearly transmit the message. In other words rather than verbalizing one philosophy and demonstrating another, the administrator should take every opportunity to act out his stated philosophy. By so doing, he transmits the sincerity of his commitment.

To the administrator his role in supervising innovative programs is a perplexing one. Often he does not understand the nature of proper supervisory behavior because in innovative programs the teacher's role is drastically altered, especially in the systems approach to instruction. What should a teacher be doing while students are engaged in CAI, programmed instruction, or self-instruction in an individual study carrel? The evaluate them. To do so would be to acknowledge weaknesses which question,[17] probably because he fails to recognize that the system rather than the teacher is the unit of instruction.

Under any plan the supervisor's task is the facilitation of instruction; it should never be the *evaluation* of instruction. Evaluation implies an adversary system—that the purpose of supervision is to assess teacher fitness. Evaluation precludes the facilitation of instruction because teachers can hardly be expected to ask for assistance from someone who will later evaluate them. To do so would be to acknowledge weaknesses which might not otherwise be apparent. Evalaution, therefore, runs contrary to trust and teamwork.

Facilitation, to the exclusion of evaluation, implies a supportive supervisory attitude. When the supervisor begins to see his role as a facilitator of instruction rather than as an evaluator, teachers begin to sense the sincerity of the supervisor's plea for teamwork and of his interest in them as professional persons. When this occurs, teachers become more open to suggestions and important progress can be made.

A supportive administrative attitude is the key to human relations leadership. Research has shown improvement in performance over time to be greatly facilitated by supportive leadership.[18] The evidence from other fields, including child development and mental health, provides a consistent pattern.

Each person is motivated by his desire to achieve. His sense of personal self-worth must be enhanced if he is to gain satisfaction and effectiveness within a work environment. The individual likes to know that he is appreciated, that he is perceived to be making valuable if not indispensable contributions to the organization. Each person needs reinforcement from his superiors and subordinates—the principal from his superintendent and teachers, the teacher from his principal and students. If he does not receive it, his morale and eventually his performance suffers. The myth may be that the person having a strong ego and utter self-confidence does not require the occasional pat on the back, but this is indeed no more than

17. Richard O. Carlson, "Programmed Instruction: Some Unanticipated Consequences," in Glaser et al., *Studies of the Use of Programmed Instruction in the Classroom* (Learning Research and Development Center, Pittsburgh, Pa., 1966), pp. 121-131.

18. The works of Rensis Likert and Robert H. Guest thoroughly review this literature. See for example, Robert H. Guest, *Organizational Change: The Effect of Successful Leadership* (Homewood, Illinois: Dorsey Press, 1962).

a myth. Each person needs reinforcement, the egocentric personality probably more than anyone.

Teachers crave approval because of basic insecurities. Generically speaking, teachers as a group lack security partially because they have not known success to the extent enjoyed by most other professionals. Part of this is due to the nature of their professional reward system but there are also other reasons. Prospective teachers are at the bottom of all college groups in terms of their intellectual abilities and as a result they do not often enjoy a high degree of collegiate academic success. Their general insecurity is manifested in a strong sensitivity to feedback from superiors and anxiety about personal competencies. As a result, most teachers are reluctant to invite observers into their classrooms and are extremely anxious when they cannot avoid observation. Visits by supervisors are especially traumatic occasions and most accept student teachers and aides only if the work relief is substantial. Many steadfastly refuse any assistance of this kind due to their lack of confidence in their professional abilities.

The effective leader has been shown to be one who is sensitive to the needs of others while not being personally sensitive about his own affairs. Th coupling of this realization with the knowledge of teacher insecurity dictates a supportive leadership style on the part of administrators. The administrator's task is gaining teacher confidence, and this can only occur in the absence of negative criticism. The successful administrator, in this regard, is easily recognized by his easy-going, non-threatening style and by the reactions of teachers who solicit his advice and who show few visible signs of anxiety when he enters their classrooms. Teachers respond to such an administrator as a friend to whom they may go for help in surmounting daily difficulties and petty annoyances. An easy flow of communications in both directions and horizontally among teachers characterizes the school within which this administrative style predominates.

A supportive administrative style is desirable under all conditions, but where an innovative attitude is desired, it is absolutely essential. Innovation demands the willingness to take a chance, to risk failure without fears of recrimination. Innovation requires an openness of mind, confidence in one's own abilities, and in the steadfast support of superiors. One can survive by cautious, traditional teaching behavior in any system—authoritarian or open; but where progressive teacher behavior is desired. an open, supportive attitude is required of educational leaders. Only with open leadership behavior can teachers be expected to risk a venture from the status quo.

The administrator of innovative programs must also provide certain logistical supports—some physical and others less tangible. The physical support may include things as costly and difficult to obtain as new, flexibly designed buildings or a new computer. But equally important it may include such simple things as cooperation by the custodial staff or the existence of a petty cash fund. It is a marvel of our time that in many schools the custodian still prescribes the arrangement of desks and chairs, arrangements absolutely essential to such breaks with tradition as group work, the discussion method as opposed to the lecture, and individualized instruction. Equally anachronistic is the willingness of many over-cautious administrators who refuse to allow a petty cash fund of a few dollars for purchase of miscellaneous, day-to-day supplies—this in the context of total budgets that may run into the hundreds of million of dollars.

A less physical, although highly tangible, logistical support to innovation and the systems approach is the setting of differential salary schedules. An essential companion to the concept of differential staffing and to the need for the continuous personal good-setting discussed at length above, the differential salary schedule is a natural outgrowth of the recognition that division of labor is required for organizational vitality. Differential roles demand varying degrees of experience and expertise which should be suitably rewarded in monetary as well as psychological terms. It is not enough to acknowledge superior performance through assignment of additional responsibilities; conventional rewards are required as well. Only when salaries are commensurate to the responsibilities is a true incentive system operational.

Under the innovation ethos another support mechanism—continuing or in-service education—is constantly demanded. The change-oriented, progressive organization cannot rest on its laurels once a particular innovation has been adapted. Implementation requires immediate technical knowledge and constant up-dating, but even more necessary is the constant reminder that the innovations of today are the traditions of tomorrow. How easy it is to fall back into the doldrums of the status quo, resting comfortable on the knowledge that one has done his bit for innovation and that it is up to others to catch up. What better excuse for disregarding the ever-evolving innovations of the present than the smug self-assurance that all obligations to change have been fulfilled. Such is the constant challenge to the administrative leader.

Selected References

Carlson, Richard O. "Programmed Instruction: Some Unanticipated Consequences." In Glasser et al. *Studies of the Use of Programmed Instruction in the Classroom.* Pittsburgh, Pennsylvania: Learning Resource Development Center, 1966.

Carpenter, C. R. *AV Communication Review,* Vol. 1, 2, 11, and 14, (1953, 1954, 1963, and 1966).

Committee for Economic Development. "Four Imperatives for the Schools." *Innovation in Education: New Directions for the American School,* 1968.

———. "Judging the Value of the Technical Resources." *Innovations in Education,* 1968.

———. "The Goals of Instruction." *Innovations in Education,* 1968.

Glasser, Robert. "The Design and Program of Instruction." *Challenge of Innovation,* 1968.

Hanson, Lincoln, and Komaski, Kenneth. "School Use of Programmed Instruction." In Robert Glasser, ed. *Teaching Machines and Programmed Learning II:Data and Directions.* Washington, D.C.: National Education Association, 1965.

Guest, Robert. *Organizational Change: The Effect of Successful Leadership.* Homewood, Illinois: Dorsey Press, 1962.

Hope, Thomas W. "Nontheatrical Films." *Journal of the Society of Motion Picture and Television Engineers* 73-76 (1964-1966).

Jackson, Philip W. "Technology and the Teacher." *The Schools and the Challenge of Innovations.* Committee for Economic Development, 1971.

Magner, Robert F. *Preparing Instructional Objectives.* Palo Alto, California: Fearson Publishers, 1962.

Meyer, James A. "Teaming A First Step for Interdisciplinary Teaching." *The Clearing House,* March 1969.

Schon, Donald A. *Beyond The Stable State.* London: Temple Smith, 1971.

Schramm, Wilbur. "Instructional Television Here and Abroad." *The Schools and the Challenge of Innovation.* Committee for Economic Development, 1969.

Stolurow, Lawrence M. "Computer-Assisted Instruction." *The Schools and the Challenge of Innovation.* Committee for Economic Development, 1969.

Suppes, Patrick. "The Computer and Excellence." *Saturday Review,* 14 January 1967.

7

JAMES O. WILLIAMS

Managing the Instructional Team: Personnel Administration

A major task within the overall realm of school administration is that of bringing to bear all the available human resources on the effort to achieve educational objectives. This task is often referred to as personnel administration. It is this facet of educational administration that is being addressed in this chapter. It is not the author's purpose to detail a specific plan for administering the personnel program in a school or school system. Such a chore is too tedious for a chapter of this nature. Instead, the focus here will be upon the establishment of the parameters affecting personnel management, a brief discussion of the social setting of the school system (within which personnel operate), an analysis of the components of a rational personnel program, a philosophic framework for establishing relationships between management and instructional staff and an analysis of the trends and issues in school personnel management.

PARAMETERS AFFECTING PERSONNEL ADMINISTRATION

Personnel administration must be viewed within the total context of administration to assume its proper importance. School management might be roughly defined as those processes and practices necessary for bringing to bear, on educational objectives, all the human and non-human resources to produce changed behavior on the part of the learner in the schools. The administrative process may normally include the tasks of planning, organizing, staffing, directing, coordinating, controlling and appraising. Obviously, the mangement of school instructional personnel is a major part of each of these tasks. It is through the proper management of the instructional staff that learning takes place in the classroom. Any school administrator who loses sight of this fact has a distorted perspective of the way things get accomplished in the educational system.

Let us look briefly at the concept, "system." Griffiths has defined a system as follows:

"A system is simply defined as a complex of elements in interaction. Systems may be open or closed. An open system is related to and exchanges matter with its environment, while a closed system is not related to nor does it exchange matter with its environment. . . . All systems except the smallest have sub-systems and all but the largest have supra-systems, which are their environment."[1]

Assuming such a definition as a legitimate one for viewing personnel management, let us then define some of the major elements impacting upon performance of instructional personnel in the school.

American Society

Public schools in American society are agencies of government. The federal constitution says nothing about education but through legal implications of the tenth amendment it has been traditionally assumed that public education is a responsibiliy of state governments. States have chosen to delegate that responsibility to local boards of education which the state legislatures have created for the purpose of governing the local schools. This, then, places the local schools to a large extent, in the hands of the people. Local boards, representing the people, along with state government have imposed certain restraints on the operation of schools.

Obviously the schools are impacted upon by various events in society. Roald Campbell has suggested that educational policy results when basic social, economic or technological forces in society are followed sequentially by antecedent movements, political activity and formal enactment of laws, regulations, guidelines, and board policy.[2]

Kimbrough has further suggested, and this supports Campbell's thesis, that educational policy is more nearly a result of the informal structure than of the formal decision structure.[3] These works are mentioned simply to lend credence to the notion that school systems are simply sub-

---

1. Daniel E. Griffiths, ed., "Behavioral Science and Educational Administration." *Sixty-third Yearbook of the National Society for the Study of Education* (Chicago: University of Chicago Press, 1964), p. 116.

2. Roald F. Campbell, "Processes of Policy Making Within Structures of Educational Government: As Viewed by the Educator," *Government of Public Education for Adequate Policy Making*, ed. William P. McLure and Van Miller (Urbana, Illinois: University of Illinois, 1960), pp. 59-76.

3. Ralph B. Kimbrough, *Political Power and Educational Decision Making* (Chicago: Rand McNally and Company, 1964).

systems of the larger society and cannot operate in isolation from society. This implies that in the administration of school personnel programs, inputs from the supra-system as well as sub-systems within the educational system are at work.

School administrators must be sensitive to the events in the larger society which impact upon personnel relations in the school. For instance, in a society in which people at all levels are enjoying a greater voice in decisions which affect them, it is impossible for school administrators to ignore the voice of the teaching force collectively or of teachers individually. This should not be interpreted to imply that the administrator should bend with every suggestion by personnel. It does imply, however, that in the long range establishment of personnel policy, teachers should have a major voice.

Governmental Structure

As was suggested in the above statement regarding society, education is a function of government. That is to say there is a formal structure through which the wishes of society must eventually be legitimized. The federal constitution allows states to formalize a structure for the operation of schools. Each state determines how schools are to be financed, how they are to be governed and to a large extent, the parameters within which they are to be administered. More specifically the state may determine the basic salary scale to be paid from state funds, certification requirements and what is to be taught. So, relative to the personnel program in a local school or school system there are certain limits regarding who is to teach, what is to be taught and how much remuneration can be allowed. To the open-minded school administrator it might appear at first glance that too many restrictions are placed upon the administration of the professional personnel program but upon closer examination these are only broad limits which allow the local boards of education a great deal of freedom to go beyond minimum requirements.

The local administrator must operate within policies which have been determined by the local board of education. Probably the major negative factor related to the necessity of operating within established personnel policy is that administrators generally accept minimal policy as being the ultimate and through a lack of creativity they never seem to see alternatives to what may currently exist. Suffice it to say, however, that the legal governmental structure does impose broad limits to which the administrator must adhere.

Bureaucratic Organization

Early man found that some tasks can be achieved through organized group effort that cannot be achieved through individual effort. Before man developed the tools necessary for securing meat he found that through cooperative efforts of individuals, a human circle could be created and progressively tightened to force a herd of animals off a cliff. The animals could then be used for food. In all of man's activity throughout history, there has been an effort to organize in ways perceived to enable man to make his life more enjoyable with less effort being devoted to that which is necessary for survival.

Merton, in discussing a rational organization has suggested that in such an organization, all activity, ideally, is functionally related to the purposes of the organization.[4] The organziation is characterized by clearly defined roles, regulations which are impersonal in nature, and strict control through authority which resides in positions rather than in the individuals holding the positions.

The ideal type of such an oganization is the bureaucracy which is characterized by the following:

1. Clear-cut division of activities inherent in the office;
2. Assignment of roles on the basis of technical qualifications;
3. Problems are categorized and dealt with in compliance with established policy;
4. The pure bureaucrat is appointed although some of the higher positions are elective and assumed to be representative choice;
5. Technical procedures for affecting the organization are performed by continuous bureaucratic personnel;
6. Maximized vocational security based on incremental salaries, promotion by seniority, and pensions.

According to Merton, the basic merit of bureaucracy is its technical efficiency. This type of organization is one which nearly approaches the elimination of personal relationships and decision making based on emotions of individuals within the organization. This has led, in the bureaucratic organization, to such commitment to discipline or predictability of performance that strict adherence to the regulations becomes an end within itself. This concept has been referred to, by various authors, as "trained incapacity," "occupational psychosis," and "professional deformation."

The career pattern of the bureaucrat is such that the person who adheres to the rules is likely to be promoted whenever he is eligible and positions are available. This tends to cause adherence to rules and strict

---

4. Robert K. Merton, "Bureaucratic Structure and Personality," *Organizations: Structure and Behavior*, ed. Joseph A. Litterer (New York: John Wiley and Sons, Inc., 1963), pp. 373-380.

discipline to actually become a part of the personality of the bureaucrat. This in turn sometimes causes tension between the bureaucrat and the public. On the other hand, a lack of adherence to the regulations may cause tension between the employee and the administration. The bureaucrat is, therefore, under pressure from the organization to adhere to regulations which are impersonal in nature and at the same time he must deal with clients who feel a need for personal treatment. Within the organization, personal consideration is frowned upon.

Many school systems believe, and justifiably so, that regulations must be very concise, designed to cover most situations, and strictly enforced. Rules makers sometimes make the rules so that compliance produces efficiency in general but promotes inefficiency in specific cases. It is impossible for rules makers to anticipate all situations which might arise.

Merton's discussion of the characteristics of bureaucracy provides considerable insight regarding organizational behavior of the school system. It is helpful to realize that while bureaucracy has the very positive characteristics of technical efficiency it may produce inefficiency in specific cases. Gouldner,[5] more clearly than Merton, points out that the tension resulting from strict adherence to regulations causes further adherence to regulations which ,in turn causes more tension. This can be seen over and over in the school setting.

It should not be assumed, however, that bureaucracy should be abandoned for some other form of organization in the school. It seems that bureaucracy may be the most efficient organizational pattern for working with large groups of people. It also has the positive aspect of defense against negative personality involvement. School administrators must, however, be aware that the instructional program can be damaged if regulations become an end rather than a means to fulfillment of the educational goals. Professional personnel must maintain (or be given) the right (and encouragement) to undertake bold new approaches to the teaching process. Bureaucratic administration must be a means to learning and not an end to itself.

Human Needs

The administrator must not lose sight of the fact that in administering school personnel he is working, not with precise machinery but with human beings. Humans have emotions, needs, motives and individual drives. Administrators can't change this no matter how they try. They can accept individuality and capitalize upon it or they can allow it to conflict with the demands of the bureaucratic organization.

5. A. W. Gouldner, *Patterns of Industrial Bureaucracy* (Glencoe, Ill.: Free Press, 1954).

It has been determined by the various psychologists who have studied human behavior that existing in each human being there is a kind of hierarchy of needs. The most basic of these is the need to survive followed by the need for security, the need for acceptance, the need for esteem and the need for self-actualization. Theories suggest that all of these needs exist and any one may be dominant at a given point in time when all of those beneath it in the hierarchy have been satisfied. It is further suggested that these needs govern man's behavior to the point that they may dominate custom, mores, laws or institutional expectations. This implies that the extent to which an individual is aware of one of these basic needs, his behavior will be influenced by the drive to satiate that need.

Basic individual needs are often in conflict with the demands of the organization. While the demands of the organization may require the individual to be dependent and passive, his mature human needs to be independent and self-directed may be more dominant than his desire to fulfill institutional expectations. There are alternative ways for the administrator to respond to individual human needs. First, he may take the position that individual needs and individual treatment of employees have no place in the educational bureaucracy. Indeed, the bureaucratic organization is designed to eliminate personal considerations. If this alternative is taken, the likely consequence is low morale, high turnover rates and a low level of commitment to the organization on the part of employees.

A second alternative is to allow individual needs to take precedence over achievement of organizational goals. A possible outcome of this is a relatively happy faculty with no assurance that the job is being done and little concern as to whether it is done. For obvious reasons, such an extreme may be highly undesirable. In the current age of accountability this could not be tolerated in the educational organization.

A third alternative is for the administrator not to assume that achievement of organizational goals and satisfaction of individual needs are mutually exclusive. Studies of satisfaction in business, industry and education have indicated that employees often express a high degree of satisfaction when they feel they are contributing, in a significant way, to the accomplishment of the tasks to be achieved. In short, we are saying that the work itself may be among the more significant factors relating to the individual's need for esteem.

The role of the personnel administrator (meaning all those who work in a superordinate capacity with professional personnel) is to recognize the existence of both the organizational expectation of people and the individual needs which impact so significantly on behavior. Only when he does this, can he effectively mediate the conflicting forces impacting upon people in the school organization. One of the positive aspects of working with professional personnel such as teachers, is that their individual needs often are evident at a level which involves self-esteem and self-enhancement rather than the lower orders of human needs. For the truly professional educator it is practically impossible for individual needs to be fulfilled in the absence of a high level of performance with relation to the professional standards commonly accepted in the teaching profession. The author is not so naive to believe that all teachers necessarily operate at the professional level. The assumption must be, however, that they do, at least until individuals prove that they do not. These individuals must then be reckoned with in the attempt to bring about the desired level of performance.

Capabilities of Students

In the industrial setting no assumption is made that the completed product is unrelated to the material provided for its assembly. Steel and aluminium are not treated the same in the manufacture of useable products. Nor is the same product manufactured from the two types of materials. We, in education, have too often assumed that every student entering the system is of the same basic material. We have also erroneously assumed that whatever material we begin with we must treat it in a "standard" way and that the product at the end of the educational process will be a standard product. We proclaim this quite dramatically by awarding a standard degree. In reality, nothing could be more inaccurate.

Our lack of control over the type of material with which we begin the educational process has certain implications for managing the instructional team. There are at least two major implications which must be considered by personnel administrators. First, in the evaluation of performance, it must be realized that teachers working with one type of student, with one level of capability, will not produce the same product as another instructional team working with students of different capabilities. The assumption must be made that the abilities and creativeness of the instructional team will determine, to a large extent, the product developed from the raw material. It must be realized, however, that students do have certain limiting characteristics.

Second, contrary to popular practice, there is a need to expose the culturally and educationally disadvantaged student to the strongest teachers available. Serious consideration should be given to the improvement of incentives for teaching in the most difficult schools. No doubt, such a suggestion will conjure up a great deal of criticism. It does, however, merit careful examination by those responsible for administering school personnel. It is both morally and legally questionable to place the beginning teacher and the one nobody wants in the most difficult teaching environment. In so doing, we are placing students who need the most help with teachers who need help in performing at an acceptable level in the profession.

The above mentioned factors are not an exhaustive list of factors which determine the framework within which personnel administration must operate. These factors are, however, influential in the total context of "people" in the school organization. Let us now turn to an examination of the various components of an adequate personnel program in the school setting.

COMPONENTS OF THE SCHOOL PERSONNEL PROGRAM

Within the context of a systems approach to school administration it seems only logical to examine the total personnel program in this chapter. To begin with, the personnel administrator must concern himself with three major periods or phases in the total career pattern. First, there must be some concern related to the *pre-employment* period. This usually involves the training of the professional person as well as the recruitment of personnel to the school or school system. It is not being implied that the personnel administrator, in the broad sense, has total control over all aspects of the pre-employment phase. School administrators can, however, make a valuable contribution to training of teachers.

Second, the personnel administrator (all of them in the system) must be constantly attentive to the *employment phase* of the personnel program. Too often, the administrator views concern for staff assignment, employee benefits, salary schedules and in-service training as comprising the entire personnel program during the employment phase. The effective personnel administrator is also attentive to morale, communication factors, effects of bureaucracy on professional personnel, alternative staffing patterns, incentive systems and evaluation.

Third, an often neglected phase of personnel administration is the *post-employment* period. Too few systems have any contact with past employees. With the continuously longer life expectancy, business, industry, military and educational organizations must develop programs to assist past employees and also to capitalize on the reservoir of experience they possess.

An attempt is made in the following paragraphs to examine some of the considerations which should impact upon the total personnel program. Let us now examine each phase suggested above in more detail.

Pre-employment

The objective of school personnel administrators during the pre-employment phase is to assure the availability of well qualifed instructional personnel and to recruit the best ones to the system. Most school personnel programs are inconsistent regarding the pre-employment phase. Some school systems, usually located in college or university towns, get actively involved in the pre-service training of teachers. This often includes the supervision of student interns. Too often, this is the extent of the involvement of school systems in the training of teachers.

The day has long past when teachers can be prepared in the "ivory tower" atmosphere of the college or university classroom. This is not to say that the college class is not valuable. It is in the classroom that the teacher trainee achieves the subject matter competence and the foundations necessary for good teaching. This is, however, only part of an effective teacher preparation program. Just as it is impossible to mediate the experience of driving an automobile it is impossible to train teachers in the absence of young people. It is one thing to study growth and development of children but quite another to be able to work within its context in the classroom. One may read that children from disadvantaged homes come to school without breakfast but be quite disturbed at the face of a hungry child in the classroom. One might also have an adequate grasp of Shakespeare and methods of teaching but be unable to communicate in a meaningful way with students in the classroom.

What is being suggested is that neither the colleges and universities nor the schools can adequately prepare teachers alone. There is a need in the preparation of teachers for colleges and school districts to work together to provide *realistic* training programs. It appears entirely feasible that the teacher trainee may eventually become a member of the typical instructional team. It is past time that all educators realize the mutual benefits of cooperation between school districts and teacher training institutions. It would appear to be a justifiable expenditure for a school district to allocate a portion of its budget to personnel employed by the school district to coordinate the efforts of interns toward the objective of providing more assistance to them during a critical stage in the pre-teaching program.

It is not at all unlikely that teacher preparation programs in the future will allow as much as twenty or twenty-five per cent of the total training program to take place in the public school setting. Many preparation programs already require some type of field experience in each of the four years of training. Such programs will require effort on the part of personnel administrators but will in the long term provide better beginning teachers.

Still another way in which school district personnel administrators can participate more effectively in pre-service training is to make provisions for personnel to participate in the formal instruction of teacher trainees. Time should be provided for the strongest teachers to go on the college campus and teach students preparing for teaching careers. Arrangements might be developed so that a public school teacher and a university professor might exchange responsibilities for a period of three months. No doubt, all concerned could benefit from such an arrangement. This would be a method of producing contact between the university professor and students in the elementary or secondary schools. Such contact is too often lacking. There are, however, scattered universities and school districts which have such arrangements.

Recruiting efforts in the pre-employment period vary widely from school district to school district and in institutions of higher education. Those systems which have had to employ large numbers of teachers have had the most sophisticated recruiting programs. Recruiting has often been viewed as a central office function with a personnel director in charge of locating, interviewing, employing and placing teachers within the system. It is the contention of the author that while this type of personnel program may be administratively efficient it is inconsistent with the need to employ professional personnel in specific situations where their unique talents may be most effectively utilized and where their philosophy may be compatible with that of their peer teachers and the school administration.

It appears logical that certain aspects of the recruiting program can be more effectively handled at the central office level. For instance, the location of available sources of teachers, compilation of personnel folders on prospective teachers, preliminary screening of credentials and determination of the overall needs of the school district should be accomplished in the central school district office. A process should be established and operative whereby a school administrator having a vacancy on the faculty could contact personnel in the central office and have them begin the search for a replacement. The personnel director should then provide the school administrator with credentials on all available candidates.

At this point it appears logical that the local school administrator should assume responsibility for the determination of a candidate to whom the position should be offered. It is in the selection process that school districts are too often in error. It should be the local school administrator, working with the faculty in the department, grade, or subject area in which the vacancy exists, who recommends to the chief school administrator that a candidate be employed. It should be clearly understood at this point that the author subscribes to the position that *selection* of personnel is a local school responsibility. While this may be more time consuming than central office selection and placement, greater satisfaction of the entire staff usually results.

Once the selection process has been completed it should then become the responsibiilty of the central office to extend a formal offer of employment to the candidate and to process employment upon the acceptance of a position by the candidate. There is an underlying principle in such a rationale. It is that those aspects of recruiting which are essentially administrative should be accomplished at the central office level while those aspects related to the professional performance of the teacher should be accomplished at the local school level. There should be an opportunity for both the local administrator and members of the probable peer group to raise questions regarding philosophy, teaching methods and knowledge of subject prior to the time of employment. It seems only logical that those most competent to judge one's probability of success in the subject or grade level of the vacant position be asked to do so.

Most of what has been said regarding recruiting has been, more or less, oriented to the large school district. In smaller school districts the major part of the recruiting task may be assumed by the local school. This may be more efficient, as well as effective, in schools where no more than two or three teachers would be employed in the course of a school year. The major point is that the people working with the teacher to be employed should be involved in the selection process. It also seems appropriate for teachers to be involved in a representative way in the selection of administrators. This is seldom done at present but such procedures are likely to become more prominent as teachers gain a greater voice in making decisions. Regardless of whether teachers demand such a voice it appears logical that teachers should have a voice in selection of administrators.

Employment Period

Various aspects of the personnel program related to the employment period normally enjoy the greatest attention from administrators. It is during the employment period that professional employees are physically a part of the system. It is during this period that their needs are most pressing, their problems are most meaningful and their demands are most threatening. Let's examine some of the major factors with which the administrator must be concerned during the employment period.

In order to most effectively utilize human resources in any organization, the administrator must find ways to release the creative talents of personnel engaged in the achievement of organizational goals. A major factor, then, with which administrators must be concerned is morale. Moore has suggested that any operational definition of morale must be based on feelings, opinions, attitudes and behavior and will, broadly speaking, be concerned with the satisfactions and dissatisfactions of the working situation.[6] There has been considerable research in recent years

---

6. Harold E. Moore, *The Administration of Public School Personnel* (New York: The Center for Applied Research in Education, 1966), p. 77.

to determine just what factors in the working environment are most prominent in promoting high morale among all types of workers. One of the more significant studies related to satisfaction—dissatisfaction was undertaken by Herzberg and others in the Pittsburgh area.[7] One of the major findings in these studies was that among two hundred engineers and accountants there were two clearly identifiable sets of factors which motivated people in the organization. One set of factors were identified as hygiene factors and were related to the job environment. These included policies and administration, supervision, working conditions, interpersonal relations, money, status and security. These factors were usually associated in one way or another with dissatisfaction as expressed by the subjects in the study. The other set of factors were motivators and were related to the nature of the job itself. Motivating factors were achievement, challenging work, increased responsibility, and growth and development. These were those factors which were related to periods in the career in which the workers were highly satisfied. A major point is that satisfiers and dissatisfiers may be on a different continuum. One therefore, cannot necessarily assume that satisfaction will occur in the absence of dissatisfying factors.

A number of studies have been conducted in education to determine the causes of satisfaction-dissatisfaction among teachers. Several of the more significant ones conducted since 1950 which examined various aspects of the morale problem were done by Bates,[8] Clark,[9] Stewart,[10] Bidwell,[11] Jones,[12] Rettig and Posamanick,[13] Butler,[14] Suehr,[15] Johnson,[16]

---

7. F. Herzberg, B. Mausnor, and Barbara Snyderman, *The Motivation to Work*, 2nd. ed. (New York: Wiley, 1959).

8. D. M. Bates, "The Morale of Teachers in Public High Schools" (Ph.D. dissertation, Dept. of Education, University of Chicago, 1950).

9. D. L. Clark, "The Dissatisfied Teacher," *New York State Education* (March 1953).

10. L. H. Stewart, "Certain Factors Related to the Occupational Choices of a Group of Experienced Teachers," *Peabody Journal of Education,* 33 (January 1956): 235-239.

11. Charles E. Bidwell, "The Administrative Role and Satisfaction in Teaching," *Journal of Educational Psychology,* 29 (September 1955).

12. J. J. Jones, "Teacher Morale and Administration," *The Clearing House,* 32 (January 1958).

13. Solomon Rettig and Benjamin Posamanick, "Status and Job Satisfaction of Public School Teachers," *School and Society,* 87 (March 1959).

14. Thomas M. Butler, "Satisfactions of Beginning Teachers," *The Clearing House,* 36 (September 1961).

15. John H. Suehr, "A Study of Morale in Education Utilizing Incomplete Sentences," *The Journal of Educational Research,* 106 (October 1962).

16. Eldon D. Johnson, "An Analysis of Factors Related to Teacher Satisfaction-Dissatisfaction" (Ed.D. dissertation, Auburn University, School of Education, 1967).

and Savage.[17] The last two were designed to see if the findings of the Herzberg study in industry would also apply to personnel in education. Savage determined that those factors which were most significant in the satisfaction of teachers generally were achievement, recognition and interpersonal relations with students. It was also determined that the manner in which they were supervised, and matters of personal life were the most significant dissatisfiers among teachers in the study. There is considerable evidence that the way a professional is supervised is a very significant morale factor.

When one looks at the normal human needs (survival, comfort, belongingness, esteem and self-actualization) it is logical to believe that the higher order of needs are the ones which are normally dominant among teachers. These needs coupled with the demands of the professional group often create stress on the teacher. A dominant need is for the teacher to operate autonomously, as a professional, and to be held in esteem by the teacher peer group. This surely has implications for the administrator whose responsibility it is to utilize human resources in the most effective way possible. Morale is not likely to be high when the supervisor behaves in ways which will not allow satisfaction of the needs of teachers.

Many studies have been conducted to determine the relationship between morale and production. To date the research has been rather inconclusive. Regardless of the lack of clear evidence that such a relationship exists, the personnel administrator may develop a sense of loyalty among teachers if there is considerable effort to allow for the fulfillment of human and professional needs.

Another major consideration during the employment period is that of communication. Too often, in the formal organization, communication is unidirectional. That is to say, communication is normally channeled from the top, downward to the teacher, but little attention may be given to communication which flows upward in the organizational hierarchy. Such communication is inadequate in the modern school organization.

To fully understand the importance of open communication in the school organization the administrator must understand the communication process which is, simply stated, the accurate transmission and reception of written or oral messages. The assumption is often made that there is high correlation between the message that is transmitted through the written or spoken word and that which is received through audio or visual senses. There are often modifiers which distort the message. To minimize the effects of such modifiers, channels of communication must be kept

17. Ralph M. Savage, "A Study of Teacher Satisfaction and Attitudes: Causes and Effects" (Ed.D. dissertation, Auburn University, School of Education, 1967).

MANAGING THE INSTRUCTIONAL TEAM **185**

open in the school organization. The following illustration may clarify this idea.

To have accurate communication there must be opportunity for feedback and clarification. In the model presented, modifiers such as the individual's feelings, attitudes and communication skills and the overall climate within the organization may act as prisms through which the transmitted message must filter. The message that is received is always

Simple Communication Model

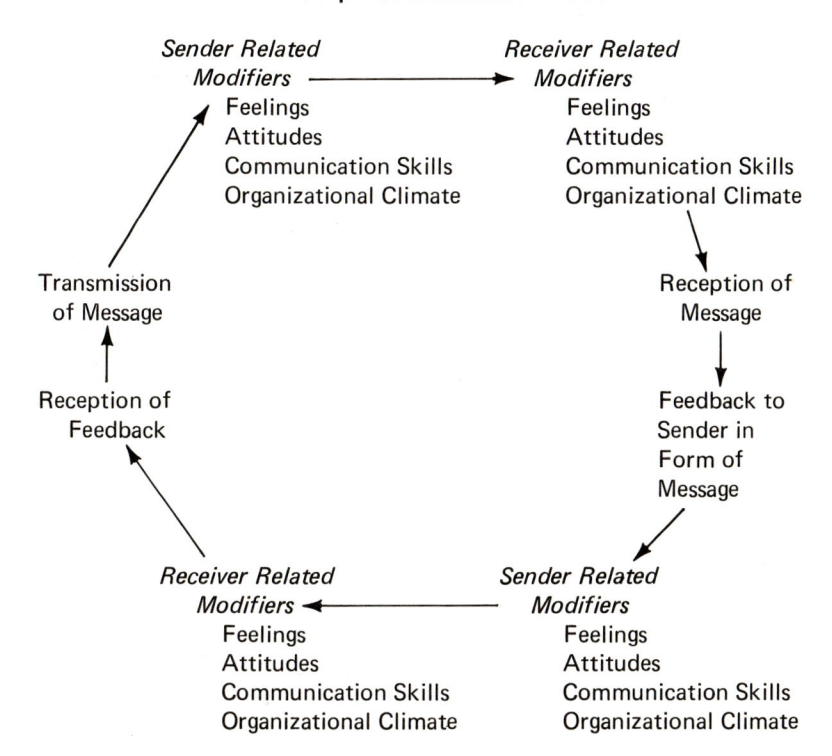

Sender Related
Modifiers ⟶ *Receiver Related*
Modifiers
    Feelings   Feelings
    Attitudes   Attitudes
    Communication Skills   Communication Skills
    Organizational Climate   Organizational Climate

Transmission of Message   Reception of Message

Reception of Feedback   Feedback to Sender in Form of Message

Receiver Related
Modifiers ⟵ *Sender Related*
Modifiers
    Feelings   Feelings
    Attitudes   Attitudes
    Communication Skills   Communication Skills
    Organizational Climate   Organizational Climate

somewhat different to that which is intended by the message sender. The modifiers acting like a prism may severely distort the message. As an example, the teacher, who may be unfamiliar with administrative terminology, may not understand a written memorandum regarding a new insurance program for professional employees. If there is opportunity for feedback the message can be clarified. If there is no feedback and clarification the misunderstanding may exist for the duration of the teacher's career or until such time as the insurance is needed. If the benefits are short of those expected by the teacher a morale problem may be created. This may then be communicated to other faculty members in distorted terms because of the feelings of the teacher who was misinformed to begin with. In short, clarification and feedback may come too late in the absence of open communication channels.

One of the positive benefits of the trend toward negotiations is that it may have a tendency to keep open the formal channels of communication. On the other hand, if negotiations take place in a threatened atmosphere or a closed organizational climate, modifiers (or barriers) may distort understanding to an even greater degree. The implications here are that the personnel administrator must constantly work at keeping the channels of communication open in all directions and in a non-threatening atmosphere. One recent approach to assure the probability of two way communication to the top of the school organization is the employment of an ombudsman. It is the responsibility of the ombudsman to provide a vital link between teachers in the school system and the administration. He is neither teacher nor adminstrator but his loyalty must be to the teacher group. At the same time, he must have the ear of the administration. One school system using the ombudsman is Dallas, Texas.[18]

Employee benefits for teachers have traditionally been somewhat short of exciting. They have, however, enjoyed some improvement in recent years. Many school districts now have group hospital and life insurance programs, disability income programs, tax-sheltered annuities, supplemental retirement programs and credit unions in addition to the usual participation in social security and state teacher retirement systems. Some improvement has come about as a result of the press by organized teacher groups to make fringe benefits more attractive.

If the administrator is to adequately provide needed benefits he must have at least two kinds of information. First, he must be informed of what the professional staff views as important regarding employee benefits. It is recommended that a welfare committee to help select employee benefit programs be operational in every school district. System wide benefit programs appear more feasible than programs which serve individual schools because of the various benefits accruing to large group membership. The welfare committee should consist of representatives of various groups in the system. For instance, it would be necessary to include both married and single teachers, younger and older teachers and representatives of all racial and ethnic groups. Obviously, all programs do not benefit all groups equally. As an example of the differential benefits to various groups, consider group medical care programs. One policy available to

18. Stephen W. Brown and Russel F. Dyer, "The Ombudsman in the Educational Hierarchy," *The Clearing House,* 46 (December 1971).

the system may be extremely desirable in terms of major medical costs. Such a policy is usually favored by middle age and older teachers. On the other hand another policy may have a low deductible figure and may be more desirable for the young male teacher with small children in the family. In such cases there may be numerous cuts, bruises or broken bones of relatively low costs while the probability of major medical care is less likely.

A second kind of information necessary for building a desirable employee benefit program is data relative to the various programs available to the school system. It is in this aspect of the personnel program that the school board attorney may be helpful in evaluating the benefits. The central personnel officer should develop files of information to be made available to the welfare committee and should himself become a student of employee benefit programs. It may often be helpful to talk with representatives of other school systems to get ideas from them.

Whenever possible, there should be a number of options available regarding insurance, annuity programs, supplemental retirement programs and credit union membership. Obviously, the larger school systems have an advantage regarding the availability of options.

In-service training and professional growth are sometimes considered by administrators to be outside the realm of personnel administration. It is the opinion of the author, however, that some dimensions of professional growth opportunities are most naturally a part of personnel administration. It would be wise at this point to draw a distinction between in-service training on one hand and professional growth opportunities outside the structure of the immediate work environment on the other hand. For the purposes of this discussion, in-service training is defined as those activities which take place in the system under the auspices of the instructional supervisory staff. Professional growth opportunity is used to imply the opportunity to study or work outside the school system in such a way that probability of promotion within the system will be enhanced. This might include advanced graduate study while employed by the school system, with all or a part of the cost paid by the school system. More and more often, school systems are providing sabbatical leave for study or travel.

Those in the school system concerned with personnel administration should continuously try to enhance the opportunity for professional growth through the budgeting of funds to promising people in the school system. This is one way that school districts can assure the development of leadership in the system. Such a practice also enhances the probability of recruiting capable people to the system.

Another requirement regarding the provision of professional growth opportunity is related to the administration of pre-service opportunity for interns, tutors, teacher aides and students participating in laboratory activities in the system. It is increasingly obvious that school systems and universities must share the responsibility for teacher training. Personnel administrators at all levels must open the doors to future teachers. While it is difficult to coordinate the use of pre-service personnel in the schools the long-range effects are mutually beneficial to the school system, the university and the teacher trainee.

Staffing patterns in schools have changed dramatically in recent years. Further change is imperative. Some school systems have moved toward differentiated staffing of the schools while others (the overwhelming majority) still assign staff to schools and classes in the traditional manner. A simple illustration may be helpful in examining some of the staffing alternatives available to administrators.

Let's assume that the average teachers' salary in a school system is $8,000 per year and the ratio of teachers to students is one teacher per 25 students. One might further assume that in the elementary school of 350 students there would be about $112,000 per year available for paying the salaries of the staff. Let's examine some alternatives.

Alternative 1—The principal of the school can proceed to assign the fourteen teachers allocated to a group of twenty-five students at a particular grade level. This has some limitation due to the fact that the number of students within a given grade level may not be a multiple of twenty-five. There might be, for instance, sixty-five first grade pupils. The decision must be made regarding whether two or three teachers will be assigned to the first grade. If we assign two teachers there will be thirty-two students in one class and thirty-three in another. If we assign three teachers to first grade, there will be two classes with twenty-two students and one class with twenty-one students. Such decisions must be made for each grade level. Staffing in this traditional way leaves us with one teacher in each classroom of approximately twenty-five students. The average salary would be about $8,000 per year within a salary schedule where the only variables are time in service and length of training. We make the assumption that teachers can do all things equally well and work with all students equally well. In fact, some teachers will be strong in one area of instruction while another teacher will be strong in another area of instruction. In most cases there is no attempt to match the needs of pupils to the strengths of teachers. We often pride ourselves in random assignment of pupils to teachers because this is "democratic."

Alternative 2—Another approach would be to differentiate staff in such a way that we assign teachers and other personnel to an instructional team and allow them to work with a larger group of students. To illustrate how this might work, let's assume that there are 130 pupils in what would normally be considered grades one and two. Going back to our teacher-pupil ratio of one teacher per twenty-five pupils, we should be allowed approximately five or six teacher units to staff this group. We can therefore assume that a staffing expenditure of $40,000-$48,000 is reasonable. Let's arbitrarily select the midpoint or $44,000. By differentiating staff we can employ the following personnel.

1 Master Teacher	$10,500
1 Professional Teacher	9,000
1 Professional Teacher	8,000
1 Beginning Teacher	6,000
1 Teaching Aide	3,150
1 Clerical Aide	3,150
2 Half-time Teacher Trainees	3,150
Specialists as Needed	1,050
Total	$44,000

Using this alternative we now have the advantage of having eight adults working with the 130 students on a continuing basis. There are, of course, other advantages to such a plan. All 130 students now have available to them all eight of the regular adults who can be called in for testing, diagnosis, prescription and special assistance. Still another advantage is the career pattern created by such an approach. As differentiated staffing becomes highly developed it may no longer be necessary for the master teacher to leave the classroom and become an administrator in order to progress in the profession. This alternative allows the most effective utilization of the talents possessed by each individual.

The two alternatives mentioned above are in no way exhaustive. There may be many alternatives within the context of differentiated staffing. Normally the high school teaching team is built around a subject area specialization but may be developed along grade lines. As more schools become non-graded there will be a tendency for the team composition to cut across normal grade lines.

Many educators, government officials, citizens and politicians are now referring to the current period in education as the age of accountability. This leads us to a discussion of evaluation of personnel. As the demand increases for educators to be held accountable for the expenditure of public money, public school personnel administrators must find more effective ways of evaluating staff. The current teacher surplus makes it even more necessary to evaluate teachers relative to their performance in service.

One of the major questions which has been debated is related to who will evaluate. In some school systems the school principal rates the teacher. In others, the supervisor prepares the evaluation. A system employed often in higher education but seldom used in elementary and secondary schools has a great deal of potential in terms of improved instruction and in terms of merit determination. That is, the peer evaluation. It seems most logical that those doing similar professional jobs are in a better position to determine merit than either the principal or supervisor.

Peer ratings appear to be extremely logical in the team approach to evaluation. When a group of several people work together in the teaching task it becomes obvious to all which people are contributing to learning and which ones are not. It may also become obvious that a person may be performing in the wrong role. Usually, the evaluations will become obvious in planning activities. Improvement will take place throughout the school year rather than at some crucial point at which the teacher may realize that his performance is unsatisfactory. The evaluation of personnel then takes on a positive flavor rather than a negative one.

In order to effectively evaluate personnel performance there must be a clear definition of realistic expectations. We have traditionally expected all teachers to possess the same skills. In reality, they do not. The teaching team should, to some extent, determine the expectations for each member of the team by specifying what each member is to do. The team can then determine the extent to which that expectation has been fulfilled.

The acceptance of any system of evaluation is dependent upon three factors. First, the teacher must be aware of what the expectations are and what criteria are to be employed in the evaluation. Second, all employees must feel that the criteria are being applied fairly by those most qualified to determine the extent to which expectations have been fulfilled. Third, there must be an incentive system which allows for rewarding those who do an outstanding job. One danger in unionization of teachers is that standard salary schedules and standard treatment are often demanded. Personnel administrators must assure that merit is recognized and rewarded in one way or another.

Post-Employment Period

Most school systems have assumed that the relationship between the teacher and the school system staff ends when the teacher ceases to be employed. This is rather unfortunate in many respects. There are several factors which favor a continuing relationship beyond the separation of the employee from the school system. Let's examine briefly some of those factors.

The teacher who has worked in a school system for a reasonable length of time may have a contribution to make to instructional improvement. An exit interview often brings to light many problems that remain undetected by administrators. A teacher may talk much more freely regarding problems after he ceases to be employed by the system. He is under no threat at this point. Such persons may prove to be valuable at a later date when the school or school system is being evaluated for accreditation purposes. They often have an unattached feeling of objectivity while they have insights resulting from having worked inside the system.

Teachers who retire from the school system have often devoted a full career to the system from which they retire. If not, they have certainly provided a long contribution to the field of education. They are, therefore, entitled to certain counseling services after their retirement. The personnel administrator at the central office level can provide valuable assistance in the initiation of retirement benefits. The need for assistance is obvious. The retired teacher is faced with tasks he has never before encountered. Such problems often occur at a time when the retiree is psychologically adjusting to a new role in life. This adjustment is often rather traumatic at best, without the added frustration of problems with social security, teacher retirement or annuity payments. The personnel office should be available to assist the retired teacher.

From another viewpoint, it seems logical that many retired teachers can contribute to instruction by assuming the new role of resource person. School systems have probably never even begun to capitalize on this valuable resource. Some states now allow teachers to retire after the completion of a given number of years in service. That person may still be very active and alert. Historically speaking, many persons have made their most impressive contributions during the years after retirement. For instance, some Presidents of the United States have been beyond the age of sixty-five. Certain Supreme Court justices have been active even after the age of eighty. There might be real value in personnel provisions which would allow retired persons to work as part-time consultants or resource persons to the school system. Such an effort could be mutually beneficial to the school system and to the retiree.

THE CHANGING CONTEXT OF PERSONNEL ADMINISTRATION

School personnel administration is undergoing change just as other phases of school administration are changing. Changes affecting the administration of personnel are not unrelated to other aspects of administration. Change must be viewed within the context of the total social system in which schools exist. The following discussion represents the most significant aspects of what the author believes to be a changing context of personnel administration.

New Concepts of Leadership

Shortly after the turn of the century scientific management became the dominant management theory in the industrial setting. This theory assumed that people were somewhat analogous to machines, could be studied scientifically and engineered in such a way that they would behave in very precise and predictable ways. Also assumed as a part of the theory was the notion that man is economically motivated and could be rewarded for his behavior in such a way that he would respond to whatever human engineering might be imposed upon him. Time and motion studies were conducted as a part of the daily routine and the worker was pushed to perform at a maximum production level. Economy and efficiency were by-words.

Eventually these ideas found their way into education. School administrators looked to the various advocates of scientific management such as Frederick W. Taylor for guidance in school management. Application of the principles of scientific management led to great concern for economy and efficiency in education with an apparent lack of recognition of the distinction between the business organization and the educational institution. Callahan was prompted to write a book entitled *Education and the Cult of Efficiency*.[19] This was a critique of the application of scientific management to education.

Later, within the first half of the twentieth century, it became apparent that man was motivated by factors other than economic ones and that there were certain social needs that must be fulfilled in order to maintain a reasonably high state of morale. The Elton Mayo studies of the Western Electric Company and the Ohio State leadership studies indicated that leadership in organization is a complex phenomenon and that man must be viewed in a social system instead of within the framework of machines. Man is, indeed, a social animal and leadership must be exerted in such a way that man's social needs can be fulfilled.

Personnel administration in school systems is subject to certain social, religious, economic and political forces operating in society. Morale in the

---

19. Raymond E. Callahan, *Education and the Cult of Efficiency* (Chicago: University of Chicago, 1962).

school is a matter of growing concern to the administrator. Public school morale in many parts of the country is at a rather low level because of such factors as poor discipline, the crisis in authority relationships, race relations, the multicultural nature of the student population, financial crises in the schools and a multitude of other pressing problems. School personnel administrators will face new challenges to develop and maintain personnel programs which utilize the most capable of human resources to meet these challenges. Encouragement of emergent leadership is a necessity in the current decade.

Humanistic Relationships in Society

There is evidence in society today of a growing humanism. Much of the humanistic movement is an antecedent of social forces in society such as the civil rights movements, anti-war attitudes, concern about existential philosophies and the demands of man for a voice in matters which affect him.

Young people today have grown up in a rather affluent American society. Those who have been participants in affluence are more concerned with humanistic motives than their predecessors because of a lack of the necessity to be concerned with economic motives. Those who have grown up in that segment of society which has been denied affluence are demanding their equal share. These factors have resulted in the obvious humanistic concern prevalent in society today.

Schools and school systems have not escaped this surge of humanism. Therefore, the personnel administrator's role will become one of protecting the interests of the professional employee as well as the interests of the school board and the public. He will be forced to build benefit programs that will improve the status of the teacher to a level comparable to that of other professions.

New Race Relations

Personnel administrators in the past have operated dual systems for black and white personnel. This is no longer acceptable just as the broader concept of dual school systems is no longer acceptable.

One of the new responsibilities now facing administrators is that of building documentary files to defend their employment of those who actually become a part of the system. Whereas several years ago it was highly unlikely that a school system would be involved in legal action relative to employment practices, it is now highly unlikely that he will escape such action. There is considerable pressure to maintain some form of balance in the number of black and white faculty members in each school. This brings to light a whole new array of problems in personnel administration. The major problem for the personnel administrator is to make sure that he is in a defensible position.

Still another related problem is assuring non-discrimination in terms of sex. There will be increasing pressure to employ women in supervisory and administrative positions at the central staff level. Many school districts have done this adequately for many years. Others are more vulnerable. Once again, it is important that documentation be provided regarding an effort to bring about racial and sexual balance in staffing at all levels in the school system.

New Legal Relationships

The legal position of public employees in all forms of public service has changed in recent years. A traditional view has held that a person employed in public service does not have the same rights as does a person employed in private business or industry. This view has changed rather significantly in recent years. There is a growing attitude that public employees cannot be denied rights afforded employees in private enterprise.

The school administrator must become increasingly aware of the legal framework within which labor-management relations exist. It is highly desirable that the school board attorney be utilized in a continuous orientation to what the relationship is between school systems and professional employees.

Professional Negotiations

Teachers, along with many other groups in society are demanding a voice in decisions which affect them. The traditional role of the teacher—meek, mild, moral, dedicated, passive, feminine—has changed during recent years. Today's new breed of teacher is a "Young Turk." He is no longer passive and feminine. Hopefully, he is committed to teaching as a career but he exercises his commitment in a different way. He may now become engaged in militant, aggressive, action to further the status of teaching as a profession. Many writers in the field have suggested that militance must precede professionalism.

One of the major factors which has influenced the teacher image is the fact that more males now enter the teaching profession. They have family responsibilities with accompanying financial obligations and are, therefore, more willing to assume an aggressive stance in improving the status of teachers. Still another factor is the entry of young black teachers into the profession. They bring with them a recent history and meaningful experience in militant action. Some of the same aggressive actions employed in their fight for civil rights may now be applied in the fight for better working conditions and a greater voice in decisions which affect them. It is obvious in some states that with the merger of the traditional black and white teacher organizations the new organization assumes a rather aggressive stance.

Personnel administrators at all levels in the school system must become students of techniques employed in negotiations. Many administrators are still threatened by the prospects of having to negotiate with teachers for salaries, working conditions and the right to make decisions. The administrator must be keenly aware of the legal parameters which govern what is negotiable. Once again, the school board attorney can be of significant value in consulting with those who negotiate.

There may be conflicting role requirements for those who administer personnel programs. It has been suggested that humanism is of growing concern and has certain implications for personnel administration. When the administrator becomes a negotiator he may be required to negotiate for the board. It is extremely important that the administrator know for whom he is negotiating.

Supply and Demand

The school personnel administrator today finds himself in a position unlike any he has ever experienced before. Traditionally, school districts have faced the task of finding bodies to place in teaching positions. For the first time in the history of American mass education there is a surplus (nationally) of certified teachers. This should not be interpreted to mean a surplus in all fields and in all school systems. Many rural school systems and all systems with vacancies in some special areas, still have difficulty in recruiting. For the first time, however, recruiting generally can be done on a selective basis.

This new recruiting situation is a relatively desirable one for the administrator but it places new demands on the school district. Instead of simply finding someone to take a position, we must now develop sophisticated recruiting techniques to assure the employment of the most qualified applicant for a position. This is no easy task and the transition from desperation recruiting to selective recruiting may be a difficult one for many administrators.

In summary, the author has suggested a philosophy of personnel administration which views the teacher as a highly important element in the system. The teacher is viewed as a professional and should be treated as one until such time as there is indication that he is not a professional educator. The problem should then be dealt with individually.

Many might take issue with the position taken by the author in this chapter because it places personnel administration in a prominent status in the overall scheme of administration. No apology is made for this position. It is the belief of the author that no single aspect of school administration is more crucial to the success of the educational system than staffing and working successfully to marshal human resources toward the accomplishment of educational objectives.

Selected References

Allen, Arthur T., and Seaburg, Dorothy I. "Principal's Role in Supervision of Pre-Service Teachers." *National Elementary Principal,* January 1968, p. 12.

Allport, Gordon W. "The Psychology of Participation." *Psychology Review* 53 (1945): 117-132.

Arnstein, George E. "The American Education Placement Service." *School and Society,* Summer 1967, pp. 298-301.

Aven, Samuel D. "Why Many Leave the Profession of Teaching." *Education,* November-December 1968, pp. 147-8.

Benson, Charles S., and Dunn, Lester A. "Employment Practices and Working Conditions." *Review of Educational Research,* June 1967.

Blau, Peter M. "Strategic Lenience and Authority." *Organizations: Structure and Behavior.* New York: John Wiley and Sons, Inc., 1963.

Buskin, M. "Problem: Dealing With An Angry Militant Teacher Organization." *North Central Assn. Q.,* January 1967, p. 76.

Cass, James, and Birnbaum, Max. "What Makes Teachers Militant? *Arizona Teacher,* March 1968, pp. 10-13.

Cohodes, Aaron. "Fort Worth Board Attacks Administrative Inbreeding." *Nation's Schools,* July 1968, p. 16.

Dawson, John. "New Approaches to Decision Making." *National Elementary Principal,* May 1968, pp. 62-69.

"Educational Evaluation: Theory and Practice." *Educational Product Report,* February 1969.

Feldvebel, Alexander M. "Teacher Satisfaction." *Clearing House,* September 1968, pp. 44-48.

Ference, Thomas P. "Can Personnel Selection Be Computerized?" *Personnel,* November-December 1968, pp. 50-56.

Festinger, L. "Informal Social Communication." *Psychology Review,* 1950, pp. 271-282.

Flinker, I. "Reporting Teacher Observation." *Clearing House,* September 1966, pp. 9-12.

Frey, Sherman H. "Policy Formulation: A Play Involving Teachers." *Clearing House,* January 1969, p. 259.

Gezi, Khalil K. "The Principal's Role In Developing Staff Morale." *High School Journal,* December 1962.

Goldman, Howey. "Conditions for Coequality." *Clearing House,* April 1969, pp. 488-491.

Grieder, Calvin. "Better Staff Leadership Can Ease Teacher Power Demands." *Nations Schools,* December 1967, p. 12.

Hare, A. Paul. "The Dimensions of Social Interaction." *Readings In Organization Theory,* 1966, p. 242.

Harrer, John M. "Superior People Are Rejecting Classroom Teaching." *N.E.A. Journal,* November 1966, pp. 20-22.

Harris, Ben M. "New Leadership and New Responsibilities for Human Involvement." *Educational Leadership,* May 1969, p. 739.

Horvat, John J. "The Nature of Teacher Power and Teacher Attitudes Toward Certain Aspects of This Power." *Theory Into Practice,* April 1968, pp. 51-56.

Horwitz, M., et al. *Motivational Effects of Alaternative Decision-Making Processes in Groups.* Urbana: University of Illinois, Bureau of Educational Research, 1953.

Howard, George W. "A Guide To Judging Women Job Candidates." *Personnel Journal,* April 1968, pp. 259-261.

Hunter, F. W. "Staffing For Variability." *Educational Leader,* March 1967, pp. 501-4.

Katz, D., and Kahn, P. L. "Human Organization and Worker Motivation." *Industrial Productivity,* 1951, pp. 146-171.

Kraft, L. E. "Rotating Principaliship." *Clearing House,* April 1967, pp. 462-464.

Kuntz, D. E. "Misassignment: A New Teacher's Burden." *Clearing House,* January 1967, pp. 271-272.

Laabs, Charles W. "The Principal as Supervisor of Instruction." *Clearing House,* December 1968, p. 198.

Lawler, Edward E., III. "Attitude Surveys and Job Performance." *Personnel Administration,* September- October 1967, p. 3.

Likert, Rensis. *New Patterns of Management.* New York: McGraw-Hill Book Co., 1961.

McKenna, Bernard H., and McKenna, Charles D. "How to Interview Teachers." *American School Board Journal,* June 1968, pp. 8-9.

McQueen, Mildred. "Are Teachers' Roles Changing?" *The Education Digest,* October 1968, pp. 8-10.

Muir, J. Douglas. "The Tough New Teacher." *American Sch. Bd. Journal,* November 1968, pp. 9-14.

Nelson, Jay L. "The Supervisors Ten Commandments." *American Vocational Journal,* February 1967.

Pino, E. C., and Johnson, L. W. "Administrative Team: A New Approach To Instructional Leadership." *Clearing House,* May 1968, pp. 520-525.

Pryor, Dayton E. "Guidelines for Successful Recruiting." *Personnel,* September-October 1967, pp. 16-21.

Rogers, Virgil M. "Appraising Teaching Efficiency for Betterment of Schools." *School Executive,* 1948, p. 54.

Romero, Ben B. "A New Pattern For Recruiting." *Personnel Journal,* April 1969.

Rutrough, James E. "The Supervisor's Role in Personnel Administration." *Educational Leadership,* 1 October 1966.

Salz, Arthur E. "Formula For Inevitable Conflict: Local vs. Professionalism." *The Clearing House,* January 1966, p. 267.

Sandberg, John "How Schoolmen Can Recruit Teachers More Effectively." *Nation's Schools,* May 1969, pp. 80-81.

Satin, Joseph. "The Principal and Individualized Teacher Supervision." *Education Age,* March-April 1969, p. 8.

Sayles, Leonard, and Strauss, George. *Human Behavior in Organizations.* New York: Prentice-Hall, 1966.

Schmuck, Richard, and Blumbery, Arthur. "Teacher Participation in Organizational Decisions." *NASSP Bulletin,* October 1969, pp. 89-105.

Sergiovanni, Thomas J. "New Evidence of Teacher Morale: A Proposal for Staff Differentiation." *The North Central Assoc. Quarterly,* Winter 1968.

Shaffer, Harry G., and Shaffer, Juliet. "Job Discrimination Against Faculty Wives." *The Education Digest,* May 1966, pp. 32-34.

Shelton, London. "Supervision of Teachers—The Administrator's Responsibility." *NADDP Bulletin,* November 1967.

Southworth, William D. "Teachers Seek Greater Voice—Teamwork is the Answer." *The American School Bd. Journal,* September 1966, pp. 65-66.

"Specialists In Elementary Education." *The Education Digest,* April 1968, pp. 8-10.

"Staff Differentiation: An Answer to Merit Rating?" *Calif. T. A. Journal,* January 1969, pp. 40-45.

Steinkamp, Stanley W. "Some Characteristics of Effective Interviewers." *Journal of Applied Psychology,* December 1966, pp. 487-492.

Stenbing, Carl M. "Some Role Conflicts As Seen By A High School Teacher." *Human Organization,* Spring 1968, pp. 41-44.

Teach, Leon. "Simulation in Recruitment Planning." *Personnel Journal,* April 1969, pp. 286-292.

Tolbert, Rodney N. "Should You Employ That Male Elementary Teacher?" *The National Elementary Principal,* February 1968, pp. 40-43.

Urich, Ted R., and Shermis, S. Samuel. "A New Role For Tired School Superintendents." *Clearing House,* January 1969, p. 294.

Vanderlip, William. "Teacher Orientation." *Clearing House,* May 1968, pp. 526-528.

Weissman, Rozanne. "Staff Differentiation." *Arizona Teacher,* March 1969, pp. 12-13.

Wiesman, Walter. "The Search For Uncommon People." *Vital Speeches of the Day,* 15 January 1969, pp. 215-218.

Worthy, James C. "Organizational Structure and Employee Morale." *Organizations: Structure and Behavior.* New York: John Wiley and Sons, Inc., 1963.

Yerkovich, R. J. "Teacher Militancy: An Analysis of Human Needs." *Clearing House,* April 1967, pp. 458-61.

Yerkovich, Raymond J. "Recruitment: Overhauling Archaic Practices." *The Clearing House,* February 1969, pp. 328-330.

8

ROBERT J. GARVUE

Financing the Educational Program

Educational decision contexts are exceedingly complex, but choices are required from among possible alternatives in the light of information, values, and consequences. A conspicuous feature of the 1960's was the cachet of new words like systems analysis or PPBES. In addition, social-science theory played a part in matters like desegregation, Head Start, and shared authority.

A reassessment of what part should be played in policy-making by both systems analysis and social science will be reassessed during the educational structural revolution which is likely to take place during the 1970's. A root question in the reassessment is the power of rationality. Those who have worked with PPBES, for example, conclude that thus far the analysts have probably done more to reveal how difficult the problems and choices are than to make the decisions easier.

The simplistic answer of spending more money for education is inadequate in most instances although it is ordinarily a component of problem solutions. A continuing weakness in gaining societal support for increased educational allocations is the lack of adequate performance measures to "prove" that additional input has paid off.

There are limits to rationality. Rationality implies the application of knowledge and reason (logic) in comparing the possible alternatives as means for reaching stated goals or objectives. Ultimate objective rationality requires a complete knowledge of the consequence of all possible alternatives and valid estimates of the occurrence of each.

It is becoming more difficult to forecast well, simply because the world is becoming more complex and more factors need to be considered. In addition, citizens are gaining more freedom for action, there are more mixes available and more trade-offs to be explored because of better information systems and because of the development of intergovernmental systems for solving problems.

Forecasting Technology

The embryonic forecasting technology is in a state of hectic growth, but tools and techniques are being developed and applied in new ways. An abstract of Adelson's description of six directions of growth follows:

1. Focus on Objectives: Program Budgeting—Program budgeting is a set of practices for allocating resources by objectives rather than by administrative functions. A problem is that the brevity of statements of objectives may result in omission of much that is important or else comprehend much that is unimportant.

2. Focus on Needed Developments: Relevance Analysis—It is essential to factor objectives into contributory sub-objectives and to be able to estimate the relevance of each based on varied points of view and information states.

3. Focus on Relationships Among Developments: Contextual Mapping—Contextual mapping is a class of techniques for arraying and displaying elements of knowledge or conjecture so that the relationships among them are evident.

4. Focus on Process Dynamics: Formal Modeling—More complex relationships can be modeled usefully as computer capabilities grow. The models can then be exercised to provide anticipations of what would ensue under varying conditions.

5. Focus on Informed Judgment: The Delphi Technique—This technique attempts to solicit and pool judgments of informed persons about future events, feeding back the results to the respondents, and allowing them to adjust or explain their judgments one or more times, based on the feedback.

6. Focus on Synthesis: Scenario Construction—Scenario construction is the process of synthesis and involves inventing credible paths between present conditions and hypothetical future states so that more meaningful choices may be made.[1]

Planners and forecasters tend to operate without adequate participation of the varied constituencies and tend to use a special language that makes little sense to the constituencies. In addition, in the drive for economic progress the legal, social, and political forces of the society have been forced to accommodate themselves to utilitarian principles. The literature includes increasing reference to economic criteria almost to the

---

1. Marvin Adelson, *The Technology of Forecasting and the Forecasting of Technology* (Santa Monica, California: System Development Corporation, April 1968), pp. 14-16.

exclusion of other criteria to be used in making decisions leading to efficiency and effectiveness.

MACRO SYSTEMS

Scarcity of Resources

It is imperative that school administrators understand both the macro and micro levels of resource allocations if they are to use the systems approach in providing fiscal leadership in educational affairs. The terms "fisc" and "budget" both refer to the revenue, expenditure, and debt activities of government and their effects on the functioning of the economic system. The administrator must know how to persuade citizens to look beyond the tax or revenue side of the fisc or budget. A comprehensive fiscal rationality concept must be applied as a guideline for efficiency in public sector decision making. Relevant criteria must include:

1. Both the revenue and expenditure sides of the budget and public sector debt.
2. All functional areas of economic activity—allocation, distribution, stabilization, and economic growth.
3. The aggregate public sector fisc or budget inclusive of all levels and units of government.
4. The fact that nonneutralities (distortions, excess burden) may be either beneficial or harmful in a system that is already operating at a "suboptimal" equilibrium.[2]

As was stated previously, the discipline or so-called dismal science of economics is being depended upon more and more in the development of rational means of determining what our priorities should be and how we should allocate our state, local, and national resources. Scarcity is the key concept of the discipline. Samuelson defined economics as:

... the study of how man and society choose, with or without the use of money, to employ scarce productive resources to produce various commodities over time and distribute them for consumption, now and in the future, among various people and groups in society.[3]

Resources available are limited in their ability to produce economic goods by both quantitative and qualitative restraints. Land (natural resources) is limited by geographical area, the magnitude of raw materials and the technology to obtain the raw material. Labor is limited by numeri-

---

2. Bernard P. Herber, *Modern Public Finance: The Study of Public Sector Economics* (Homewood, Illinois: Richard D. Irwin, Inc., 1971), p. 100.

3. Paul A. Samuelson, *Economics*, 5th ed. (New York: McGraw-Hill Book Company, Inc., 1955), p. 6.

cal size, age distribution, and qualitatively through ethical, health, and educational standards. Capital is limited by past capital formation behavior and by the relationship of capital stock to technology.

Education is a labor-intensive institution. Approximately 80 per cent of the cost of K-12 programming is in wages and salaries—not only for instruction, but for administration, maintenance and operation, and auxiliary charges. Contrasted to education is the automobile industry in which 30-35 per cent of cost is for labor. The oil industry is capital intensive in that only 5 per cent of cost is for labor.

In the private sector of the nation, capital expenditures as a percentage of the gross national product (GNP) from 1960 through 1969 show how far behind the United States has been lagging. For the period, the U. S. annual average was 13 per cent, less than half Japan's 27 per cent average and well below Germany and the Netherlands.

Claim is that our balance of payments problem, inflation problem, employment problem, and economic growth problem are really the same problem—the inadequacy of capital growth. Likewise, claim is that the educational systems must swing away from an extreme labor intensity to a greater balance between the use of labor and capital.

Much polarization in the United States and among other nations focuses upon the current choice of priorities. Americans cannot have everything in spite of a trillion dollar economy. So-called welfare economists and even certain industrialists are making the point that not only must there be a major adjustment among choices within the public sector but also a major adjustment between the private and public sectors. Currently, 72 per cent of resources are allocated via the private sector and 28 per cent through the public sector. It is being realized increasingly that we must accept a diminished private standard of living to make our public life bearable.[4] Unfortunately, too few politicians understand this or are unwilling to become political statesmen.

Many citizens are begining to wonder if we can afford tomorrow. The General Electric Company has invested a sizeable amount of its resources to forecasting its future business environment in a world context. The future international context was expected to be characterized by:

1. Maintenance of the basic international power structure;
2. Continued economic growth, despite a conflict between population growth and economic development in the emerging nations;
3. The emergence of Japan as the third major power;

4. J. Irwin Miller, "Changing Priorities: Hard Choices, New Price Tags," *Saturday Review* (23 January 1971), pp. 36-37, 78.

204 FINANCING THE EDUCATIONAL PROGRAM

4. Polarization of main world tension on a North-South ("haves" v. "have-nots") axis rather than an East-West (Communist v. Capitalist) axis.[5]

The future domestic scene can be interpreted as the interaction of eight basic trends:

1. Increasing affluence
2. Economic stabilization
3. Rising tide of education
4. Changing attitudes toward work and leisure
5. Growing interdependence of institutions
6. Emergence of the post-industrial society
7. Strengthening of pluralism and individualism
8. The Negro/urban problem[6]

It was predicted that the U.S. domestic institutions as we know them today will endure in substance, but change in function, style, and values. There appears to be little support for radicalization of the society, or even of the universities. There is a core of conservatism in the American make-up that will act to make these changes evolutionary rather than revolutionary. Nevertheless, these changes will be pronounced and widespread in all our institutions as a result both of internal pressure from the changing aspirations and value systems of their members and of external public, governmental, economic, and technological pressure.

Having discounted radicalization, one must not go to the other extreme of hypothesizing that little will change. Indeed, it is probable. . . . :

1. That technological change, by itself, will demand a great institutional change. There are three sweeping and pervasive institutional changes that technology requires of any organization and society if it is to be both fully utilized and adequately controlled:
   a. The school, for example, must encourage experiment, flexibility, and variety—part of this must come about from a re-structuring of the management system.
   b. There must be a thoroughgoing democratization of the system; science and technology, which set in motion a process of continual change, are inhibited by the traditional hierarchical and authoritarian system, but flourish in a climate that is egalitarian, pluralistic, and open to dissent.
   c. There must be created a capability of seeing technological change as a whole—not just an economic indicator like the GNP, but indicators of the social, political, cultural, and psychological state of the nation.[7]

---

5. General Electric Company, *Our Future Business Environment* (April 1968), p. 65, and *Our Future Business Environment: A Re-Evaluation* (July 1969), p. 43.
6. *Ibid.*
7. *Ibid.*

Anticipation is that all organizations will be operated less and less by the dictates of administrative convenience, more and more to meet the wants and aspirations of their membership. Managers must expect that considerations of individual motivation, group relationships and personnel costs will weigh more heavily in their future decision making. The variety of individual goals and the demand for participating in the goal-setting and implementing will not be a smooth, orderly, or totally planned process.

Economic criteria will be among those to be considered in the determination of efficiency and effectiveness. Obviously, criteria from the disciplines of political science, sociology, psychology, anthropology, history and statistics will be used also.

Education as an Economic Problem

Since economics is concerned with allocating resources, the relationship to education is logical. The task of educational institutions is to select a proper mix of resources to meet unlimited educational needs and wants of the varied clients. Thus, education can be described as an economic choice whereby educational effectiveness must be related to the use of scarce resources to achieve goals and objectives which have been judged to be in some order of priority. Again, economics or economizing is an attempt to make the most efficient use of resources available within a unique set of constraints.

The fundamental equilibrium condition that has to be satisfied if a society is spending its resources on the variously priced goods so as to make it best off in terms of utility or well-being is that every single good or service is bringing an equal marginal utility proportional to its price. This means that ideally there must be exact equality among the ratios of each good's or service's marginal utility divided by its price. Each good or service is consumed up to the point where the last unit of money spent on education, for example, provides utility equal to that of the last unit of money spent on defense, health, fire, safety, etc., in turn; i.e.,

$$\frac{MU_a}{P_a} = \frac{MU_b}{P_b} = \cdots = \frac{MU_z}{P_z}.$$

Likewise, the utility of the last unit spent on elementary education will equal the utility of the last unit spent on secondary education, and the utility of the last unit spent on English will equal the last unit spent on art, etc.

In a non-monetary way, a student must maximize his grade average by gaining the same marginal grade advantage from the last unit of time spent in each alternative use.

Allocation Systems

The two primary institutions for effecting allocation decisions are the market and the government. The market means is characterized by the forces of supply and demand and the price mechanism, based upon consumer sovereignty and by consumers choosing with dollars. The governmental means is accomplished basically through budgetary practices of taxing and spending, and by voters choosing with ballots.

Although emphasis thus far in this chapter has been upon the allocation of scarce resources, school finance as a part of public finance must be related to the other three major areas of economic activity—distribution, and economic growth. Educational decisions effected through budgeting will affect these goals. While allocation is concerned with the division of scarce resources between alternative uses and between the two economic sectors (private and public), distribution deals with the division of the society's income and wealth among the people of the society; stabilization focuses upon the macro-economic aggregates of full employment, output, and stable prices; and economic growth relates to a "satisfactory" rate of increase in a society's resource base and the subsequent growth in output on both a per capita basis and in constant dollar terms over a period of time.

The Employment Act of 1946 and educational policy must be interrelated in numerous ways if economic goals are to be reached. For example, growth in the economy, i.e., the increase of output per capita relies upon (1) increases in the stock of human capital through investments in education, training and experience; (2) increases in the stock of non-human capital through investment in equipment, machinery, and plant; and (3) improvements in the state of our scientific and managerial technology through investment in research and development, better management and organization, and more efficient production techniques.

Economic Man v. Social Man

Eastburn has conceptualized the conflict between economic man and social man.[8] Each of us, of course, is both an economic and a social man. However, those who are 90 per cent economic man see today's world differently from those who are 90 per cent social man. Let us consider the following shorthand description of characteristics and concerns:

---

8. David P. Eastburn, *Business Review* (Federal Reserve Bank of Philadelphia, October 1970), p. 3.

Economic Man	Social Man
Production	Distribution
Quantity	Quality
Goods and Services	People
Money Values	Human Values
Work and Discipline	Self-realization
Competition	Cooperation
Laissez-faire	Involvement
Inflation	Unemployment

Economic man tends to be concerned primarily with producing goods and services and with quantitative problems. He believes that a relatively free pursuit of self-interest has served this nation well. Social man sees the good life reached by a better distribution of wealth and income. People rather than things are stressed as is freedom not to pursue one's self-interest but to realize one's true individuality by involvement in a cooperative way in solving society's problems. In education, the professor is an example of economic man. He competes at a national level with his texts and is higher paid than is the elementary teacher who emphasizes cooperation rather than competition.

Economic man has been asking authorities not to let up on efforts to curb inflation until it is licked; if this means recession, the admonition is that we'd better pay the price now than a bigger one later. Social man fears that a recession hurts most those who are already disadvantaged. When unemployment rises, as it must when the economy slows, those who are laid off first are the unskilled; efforts to recruit workers from the ghetto are suspended. Social man, therefore, is inclined to trade inflation for jobs.

Social man must convince economic man that this nation cannot prosper unless action is taken to solve social ills. It is true that unemployment compensation and minimum income maintenance provide buffers between the disadvantaged and recession. Better training and education make it possible for those who are presently disadvantaged to hold their own in recession. Social action promises economic man not only expanding markets in which to sell his wares but a more stable economy in which to produce them.

The term social balance was used by Galbraith to describe a satisfactory relationship between the supply of privately produced goods and those of the public sector.[9] Taxes in the United States represent about 28

---

9. John Kenneth Galbraith, *The Affluent Society* (Boston: Houghton-Mifflin Company, 1958), pp. 254-55.

to 30 per cent of the GNP placing the country in the same category—26-30 per cent of GNP— as Belgium, Denmark, Italy, Canada, and New Zealand. In a number of other countries, the public sector gets a much larger slice of the GNP—e.g., Sweden, France, West Germany, Norway, Austria, the Netherlands, and Great Britain. This group of nations spends between 33 per cent to 41 per cent of the GNP. Were the United States to duplicate the effort of Sweden, governmental revenue would be increased by approximately $100 billion annually.

The gross national product is now approximately $1.05 trillion and will soon zoom to 1.3 trillion before the celebration of the nation's 200th birthday. There is one economic circumstance which is partly responsible for a great deal of unhappiness nationwide in spite of such productivity. The public sector of the nation is starved in the non-defense category. The percentage of production spent in the public sector vacillates between 28 and 29 per cent compared with 38 per cent in the advanced Western European society.

The nation suffers from a decrepit public transportation system, ill functioning schools in numerous districts, a high infant mortality rate, and eyesores in the environment. While the crime rate soars, less is spent on all forms of public law enforcement than on crop subsidies.

In Fiscal Year (FY) 1969, approximately $29 billion was spent nationwide each for a "good time," the Vietnam War, and new automobiles, and only the same $29 billion to educate 43.7 million elementary and secondary students.[10]

Relative to the distribution function of government, personal income in the United States is like an expanding pie: it keeps getting bigger but the way it is sliced stays about the same. Since 1947, the richest fifth among U. S. families have received more than 40 per cent of the income, while the poorest fifth has received approximately five per cent. There hasn't been any effective redistribution. However, a so-called service strategy has been implemented to effect redistribution. The educational program Head Start, for example, has been provided for children of low-income families. The result is relatively more public services to the poor to compensate for relatively low income provided through the imperfect private sector. Title I of ESEA, likewise can effect a redistribution of income through an indirect route of service.

---

10. Robert J. Garvue, *Modern Public School Finance* (London: Macmillan Company, 1969), p. 324.

SHARE OF INCOME—UNITED STATES

	1947	1969
Lowest fifth	5.0%	5.6%
Second fifth	11.8	12.3
Third fifth	17.0	17.6
Fourth fifth	23.1	23.4
Highest fifth	43.0	41.0

The Gross National Product (GNP)

The gross national product is the broadest measure of the nation's economic activity. It summarizes the amount of goods and services available in dollars during a fiscal year. It provides a common denominator for the goods produced by the builder, the miner, the equipment manufacturer, and the farmer. Likewise, it provides a uniform measure for services rendered by the teacher, the medical doctor, the airlines and the theater. Unfortunately, the housewives' housekeeping work at home is ignored.

The Office of Business Economics of the U. S. Department of Commerce computes the total U. S. output in two ways: First, if one adds up the market value of goods and services or the consumer purchases, business investments, net exports, and government purchases, he would be computing the GNP. A second way is to add up all costs to produce the goods and services. Costs would include all incomes, plus indirect taxes and depreciation.

The interaction of the different parts of the economy is easier to understand because of the GNP. If the private business sector cuts down on spending for plant and equipment, the resulting slack can be taken up by government purchases of goods and services if a downturn in economic activity is to be avoided. To lower taxes could be as irresponsible as the directions in the 1930's to "save your money."

As alluded to earlier, a key question is how the GNP should be distributed. The goods and services produced and consumed are determined through a complex interplay of personal wants, business strategies, government actions, and availability of financial resources. There is no single scientific means of effecting the best allocation. Results depend a great deal upon differences of interest, value, and opinion among people. These differences are reconciled partially through political processes.

Chart 1 summarizes how the GNP was divided in 1969. Consumer demand accounted for 62 per cent, government purchases for 22 per cent, and business investment for 15 per cent. Of the $931.4 billion total, 20 per cent went for hardgoods such as refrigerators and autos, 29 per cent for

210 FINANCING THE EDUCATIONAL PROGRAM

CHART 1

WHERE THE GNP GOES

Market Value of Goods and Services in 1969—$931.4 Billion

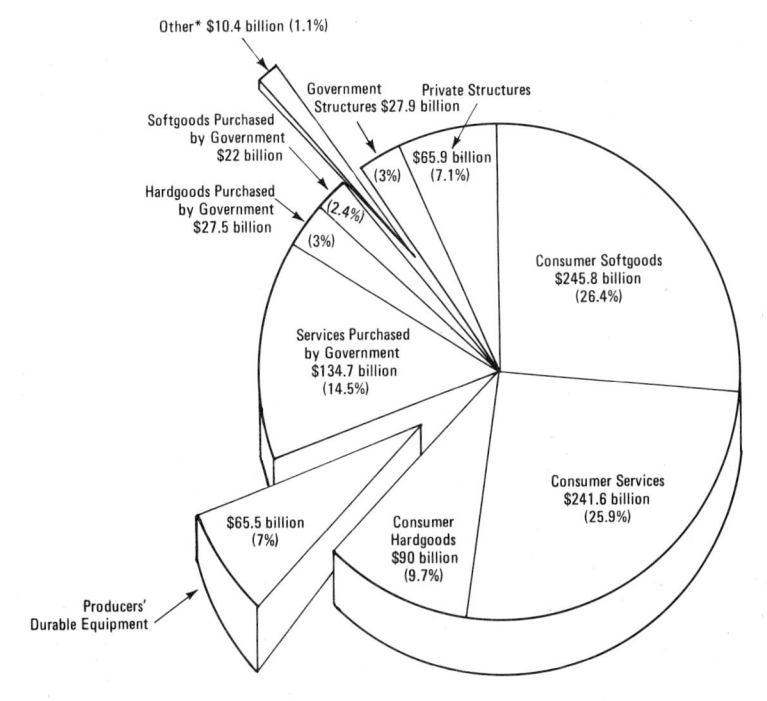

Other* $10.4 billion (1.1%)

Government Structures $27.9 billion

Private Structures $65.9 billion (7.1%)

Softgoods Purchased by Government $22 billion (3%)

Hardgoods Purchased by Government $27.5 billion (2.4%) (3%)

Services Purchased by Government $134.7 billion (14.5%)

Consumer Softgoods $245.8 billion (26.4%)

Consumer Services $241.6 billion (25.9%)

Producers' Durable Equipment $65.5 billion (7%)

Consumer Hardgoods $90 billion (9.7%)

*Includes "change in business inventories" and "net exports."
SOURCE: U.S. DEPARTMENT OF COMMERCE, OFFICE OF BUSINESS ECONOMICS

softgoods such as clothing and food, 41 per cent for services, and 10 per cent for structures.

If we want to get an idea how efficient the economy is, we can get an answer with help from the GNP. First, we have to take the inflation out of the GNP. What remain is a measure of real output—GNP adjusted to eliminate the effect of price changes. We can combine that with the total hours worked to yield one measure of economic efficiency—productivity, or, in technical terms, output per man hour. Calculations show that the hourly production of the American worker today actually is over one-third higher than ten years ago, thanks to more and better tools and skills.

The GNP can be deceptive if one doesn't account for the negative production included. In other words, some production leads to a fouling of air and water, the spoiling of scenery, all of which can lead, however, to production of other goods and services to eliminate such negative results. When a man's wife dies, the GNP will probably increase since upon the hiring of a maid the latter's salary will be added to the GNP, whereas the wife recevied no salary, and her equivalent work wasn't recorded.

Investment in education raises labor productivity. Improved labor quality has contributed 0.5 out of 2.5 per cent or as much as 20 per cent of the average annual rate of increase in labor productivity. It is realized now that an increase in the intangible capital invested in human beings and an increase in efficiency due to technological change and other factors are not less important and not less reliable than tangible capital as sources of increased labor productivity.[11]

Individuals, families, neighborhoods and society in general benefit from education. In other words, there are so-called spillover effects. Because of a decentralized system of education, local decision makers may ignore benefits or costs accruing to other geographic areas or individuals. For example, the movement of relatively uneducated persons from one state to another or from one county to another causes what is known as a spillover effect of local or provincial decision making. Each unit of government will think it is being efficient, but the totality of decisions nationwide and statewide will be inefficiency.

Educational Expenditures and the GNP

Total expenditures for public and non-public schools at all levels of education from kindergarten through graduate school amounted to approximately $70 billion during the 1969-70 school year. Educational expenditures have risen rapidly in recent years, reflecting the growth of the school age population as well as the increased efforts of the nation to provide quality education for its people. The annual expenditure is now eight times its 1949-50 total (not allowing for changes in the purchasing power of the dollar), and further increases are projected for the years just ahead.

The percentage of the gross national product which went for education has varied considerably over the past generation. Educational expenditures were relatively high in the mid-1930's, exceeding four per cent of the GNP in 1933-34. They declined sharply to 1.8 per cent of the GNP ten years later. Except for a brief period during the Korean conflict when the annual investment in education tended to stabilize, there has been a steady increase in the proportion of the gross national product spent for

---

11. Solomon Fabricant, *A Primer on Productivity* (New York: Random House, 1969), p. 52.

education ever since the end of World War II. Expenditures in 1969-70 were estimated to be at an all-time high both in terms of actual dollars and as a percentage of the gross national product (7.5 per cent). The K-12 expenditures were 4.2 per cent of the GNP.

Table 1 is a summary of data on the contributions of the three levels of government for public school financing in selected years a decade apart between 1929-30 and 1969-70.

TABLE 1

Trends in Sources of School Revenue Receipts By Level of Government
(In Millions of Current Dollars)

Year	Federal Amount	Per-cent	State Amount	Per-cent	Local Amount	Per-cent	Total Amount	Per-cent
1929-30	7	0.3	354	17.0	1,728	82.7	2,089	100.0
1939-40	40	1.8	685	30.3	1,536	67.9	2,261	100.0
1949-50	156	2.9	2,166	39.8	3,155	57.3	5,437	100.0
1959-60	649	4.4	5,766	39.1	8,332	56.5	14,737	100.0
1969-70	2,545	6.6	15,645	40.7	20,286	52.7	38,476	100.0

Source of Data: U.S. Office of Education except for the years 1969-70 which was estimated by the National Education Association.

Of the total $219 billions in total government tax revenues produced in FY 1968 for all governmental functions, the percentage of revenue collections by level was as follows:

Federal — 67.2 per cent
State — 18.0 per cent
Local — 14.8 per cent

The federal government collected two-thirds of all tax revenues, and yet revenues for public schools from that level of government were a minor 6.6 per cent.

A school administrator who understands the macro system should be able to provide much more effective leadership than one who has knowledge only of the state's minimum foundation program. Knowing that the federal government collects 67.2 per cent of tax revenues is bound to give an administrator a different perspective of the "fed's" role in school finance. Pointing out the effect of school district expenditures upon the community's economy should enlighten the local Chamber of Commerce, which might be looking at the revenue side of government to the exclusion of the expenditure side. In addition, information in Table 2 can be used very effectively to point out that education isn't bankrupting America.

TABLE 2

Percentage Distribution of GNP in Current Prices, by Function, 1955, 1966, and 1969

Function	Per cent of Total,	GNP,	Current Prices
	1955	1966	1969
Total GNP	100.0	100.0	100.0
Basic Necessities	45.7	42.3	41.6
Education & Manpower	3.7	5.7	6.3
Health	4.1	5.6	6.4
Transportation	10.6	9.9	10.0
General Government	2.0	2.7	3.1
Defense	9.3	7.8	8.3
New Housing	5.9	3.5	3.7
Business Fixed Investment	9.6	10.9	10.7
Net Exports and Inventory Change	2.0	2.7	1.1
All other	7.1	9.0	8.8

Note: Detail will not necessarily add to totals because of rounding.
Source: Department of Commerce and Council of Economic Advisers as reported in *Economic Report of the President*, February 1971, p. 99.

It should be clear that with the exception of relatively few wealthy school districts in the nation there is no way for a local school administrator to offer adequate solutions to school finance problems. In other words, school districts need equal access to state and national wealth if the concept of equal educational opportunity is to be implemented. A California court decision is likely to be the begining of a revolution in school finance which will minimize a state-local financial partnership and emphasize a state-federal financial partnership.

Serrano v. Priest

On August 30, 1971, the Supreme Court of California, by a vote of 6 to 1, upheld the complaint of John Serrano, Jr. that California's system of financing public education was in violation of the 14th Amendment of the United States Constitution, which forbids a state to "deny to any person within its jurisdiction the equal protection of the laws." Mr. Serrano first filed a complaint in 1967 along with parents of 26 other Los Angeles County schoolchildren after a school principal called him and suggested that "You've got a couple of very bright kids—get them out of East L. A. Schools if you want to give them a chance."

Mr. Serrano, a psychiatric social worker had been active in Chicano community affairs and had believed that the schools were good. Within a few months after the conversation with the principal, Mr. Serrano left the Mexican-American barrio of East Los Angeles and moved his family to a middle-class suburb of Whittier, California. Soon after moving he signed his name to a complaint.

The complaint was unsuccessful in two lower courts but the Supreme Court of California brought it up for hearing. In its decision, the high court established the principle of "fiscal neutrality"—i.e., education was to be a function of state wealth rather than of local wealth. The city of Beverly Hills taxed its citizens $2.60 per $100 of assessed valuation and had a $1,500 per child education program, while in the same county the city of West Covina taxed $4.30 per $100 and raised less than $700 per child. Under the Serrano decision, each child was to have equal access to the state's wealth with an equal effort.

The "domino effect" of Serrano proved true as federal courts in Minnesota and Texas agreed in the Fall of 1971 and the New Jersey Superior Court agreed in January, 1972, that the traditional system of school finance based mainly on local property taxes is based in favor of the rich. A nationwide campaign to knock down reliance on local property taxes has been mounted by 140 lawyers representing 23 states. Even President Nixon seemed stunned by the inequities pointed out in Serrano and repeated again the possibility of a national value added tax (VAT) as a remedy. A consensus for a revolution in school finance is forming and it represents a phase of a larger movement in the development of a national community.

The United States is moving through what has been termed a period of unusual political-economic power activity. New patterns of power are in the making and the ultimate design is not clear. Service industries are assuming political-economic powers yielded by manufacturing; conglomerates are replacing smaller and medium-sized businesses; the federal establishment collects two-thirds of the nation's tax revenues and cities and states are becoming part of a national state; the mass media is influencing more than teachers and politicians; and, unlike adults over 40, those in their 30's and younger are unwilling to settle for less than full equality and social participation.

Phase II of Nixonomics has been termed a struggle over public power affecting the distribution of income. Most of our economists and politicians are using outdated theory which relies on invalid assumptions relative to the market (private sector) and have relied on the GNP as an indicator of the health of the nation. Happiness is a rising GNP. Obviously, new economic theory is needed and more emphasis needs to be placed upon distributional equity than upon production efficiency. Serrano implementation can assist in the redistribution of income through additional educational services to the lowest fifth of the nation's population who receive only 5.6 per cent of the national income. This will generate political rhetoric about socialism, radicalism, and similar emotional terms, in spite of the acknowledged mixed economy of the United States and traditional subsidies to all economic levels of the society.

Serrano and Reform

Myers cited a list of reform possibilities opened up by Serrano: overall property tax reform, the redrawing of school districts to encourage socioeconomic and racial integration, massive extra resources to poor students, extensive research into effective education, and the ending of intradistrict discrimination of the poor within cities as well as federal intervention to end disparities among states.[12] However, she suggested that just a few months into the "post-Serrano" period, "the mood has changed from elation to extreme caution. For while Serrano may hold out the promise of the resurrection of American education, it also holds out the possibility of equity without fairness."[13] Analysis of recent data on urban school financing indicates that cities don't necessarily lack a relatively satisfactory fiscal base but rather have excessive needs. Unless state school-aid formulas include weightings for needs of urban education, Serrano may not provide the added assistance to urban schools as was anticipated. In others words, a flat grant of so many dollars per pupil—for example, $850—without provision for the funding of more costly programs for target populations such as the culturally disadvantaged, physically handicapped, vocational education, and others would be an inadequate provision.

One of the most significant consequences of Serrano ultimately could be one suggested by the National Urban Coalition. The Coalition believes that the California Court's rationale will increase "housing opportunities for racial and economic groups that are now being discriminated against." The Coalition pointed out that "one reason put forward for opposing the introduction of low-moderate and moderate-income housing in suburban areas has been the claim that such housing would negatively affect the local tax base because of increased school enrollment and costs."[14] The argument would be eliminated if adequate school financing did not depend on the local property tax base. The political implications of such homogenization are staggering to the minds of perceptive politicians at all points along the continuum of social and economic philosophy.

12. Phyllis Myers, "Second Thoughts on the Serrano Case," *City* (National Urban Coalition, Winter 1971), pp. 38-41.
13. *Ibid*.
14. "Unequal Levies = Unequal Learning," *The National LW Voter* (League of Women Voters of the United States), Vol. 21, No. 5 (November 1971), p. 24.

National Disparities, Fiscal Capacity, and Effort

Current expenditure per pupil in average attendance ranged from a low of $489 in Alabama to a high of $1,429 in Alaska for Fiscal Year 1971. Controlling for the relatively high cost of living in Alaska would place the State of New York in rank one at $1,370. The average current expenditure per pupil nationwide was $839 or a cost of 71.7 cents per student hour. The latter was computed as follows:

$$\frac{\$839}{1170 \text{ hrs.}} = 71.7 \text{ cents.}$$

The 1170 hours represent 180 school days × 6.5 hours daily.

As was stated previously, the per cent of revenue received on the average from federal, state, and local sources for public elementary and secondary schools was 6.6, 40.7, and 52.7, respectively. In the past ten years, the federal government has added only 8.7 per cent of the total new revenue. During the same period, new state revenues accounted for 41.8 per cent of the new revenue for schools, and local sources accounted for 49.5 per cent of the new revenue.[15] Of the new revenue generated in FY 1971, the federal share was $125.9 million, new state revenue—1.6 billion, and new local revenue—$8.0 billion, or 3.4, 42.7, and 53.9 per cent, respectively. Spatial or geographic spillovers of education and the shortcomings of the property tax are among the several reasons dictating greater future contributions of state and federal governments.

In measuring fiscal capacity of the fifty states, a first approach can utilize economic indicators, primarily measures of income and a second approach evaluates the tax bases available within a state, and estimates the amount of revenue those bases would produce if they were subjected to various rates of taxation such as the average rate throughout the nation.

A study conducted for the National Educational Finance Project developed estimates of personal income per capita in the 50 states by deducting from total personal income (1) an allowance of $750 per person to cover basic expenditures for food, clothing, and shelter and (2) federal personal income tax paid. The net income per capita ranged from a low of $2,192 in Mississippi to a high of $4,537 in Connecticut for the year 1969. Net income per child in average daily attendance in school ranged from a low of $5,624 in Mississippi to a high of $18,772 in New York. Greater interstate variation existed in revenue capacity than existed in per capita personal income.[16] Variations in fiscal capacity among school districts

---

15. Committee on Educational Finance, *Financial Status of the Public Schools* (Washington, D.C.: National Education Association, 1971), p. 36.

16. Roe L. Johns et al., "Variations in Ability and Effort to Support Education," *Alternative Programs for Financing Education* 5 (1971): 68, 79 (Gainesville, Florida: National Educational Finance Project).

and other units of local government typically are greater than the variations in fiscal capacity which exist among states.

Most studies of the fiscal effort of school districts have employed the property tax rate or the per cent of personal income expended for public education as criteria of fiscal effort. Using the state and local revenues for schools as a per cent of net personal income in 1969, the State of New Mexico ranked first with a percentage of 8.90, while Nebraska ranked last with a percentage of 5.0.[17] The States of New York and Iowa made approximately the same effort—7 per cent of net personal income—and New York's current expenditures were $369 more per pupil. Obviously, citizens of these two states did not have equal access to the national wealth, which would be a national goal implied from Serrano.

The increasing use of non-property taxes levied by school districts has not had an equalizing effect according to reports of the National Educational Finance Project. In fact, "non-property taxes are disequalizing in that those districts which have the greatest fiscal capacity as measured by their property tax base almost invariably obtain the largest amount of revenue from non-property taxes. Thus, the use of local non-property tax levies tends to increase the revenue disparities among school districts rather than to equalize their fiscal capacity.[18]

Fiscal Capacity and Equal Educational Opportunity

There now appears to be an increasing general acceptance of the objective of equal educational opportunity within a state but major roadblocks in the immediate future lie in the way toward accomplishment of a national equalization program. Within states, the educational opportunity available to a child will not be determined by the accident of birth in a poor or a wealthy community, or in one that places a high or a low value on education. One criterion is that there should be made available to all children a minimum level of educational resources. The prime point in Serrano was that each pupil would have an equal access to educational resources of the state.

If Serrano is to be extended to effect a national or universal program of equal educational opportunity, additional costs can be approximated. With approximately 46 million pupils enrolled in the public schools in 1970, and an average current expenditure per pupil of $839, raising the level to this amount in all states in which it was less would have cost an additional $4 to $4.5 billion.[19] A more productive goal would be a mini-

17. *Ibid.*, p. 76.
18. *Ibid.*, p. 93.
19. Harvey E. Brazer, "Federal, State and Local Responsibility for Financing Education," *Economic Factors Affecting the Financing of Education*, Chapter 8, p. 253 (Gainesville, Florida: National Educational Finance Project, 1970).

mum that taxpayers of richer school districts in states with relatively generous aid systems are willing to support—perhaps $1,000 to $1,200 per pupil annually. Achievement of either level suggested at an equal or substantially equal cost to local taxpayers would require major state and federal aid programs because of the superior revenue sources most readily accessible to state and federal governments.

An assumption is that whatever the basic minimum established for per pupil expenditures, it should be subject to adjustments for differences in prices or costs and take into account the number of children in each district who suffer from varied types of deprivation. Unfortunately, there is only a limited data base of such information.

Levin has summarized recent studies which have examined the relationships between school expenditures and student achievement scores while attempting to control for differences in socio-economic backgrounds of students.[20] Ribich and Kiesling both found that additional expenditures seemed to have a greater impact on pupil achievement at low expenditure levels than at high ones. Concerning expenditure-component studies, there is a consistent finding that teacher salary levels show a positive and significant association with student achievement when other measurable influences are held constant.[21] In a study by Ribich, for example, there was a tendency for lower-status students to have greater achievement scores the higher the level of school district expenditures, but the relationship was a declining one as expenditures rose. Kiesling found that the apparent effect of additional expenditures on achievement varied according to the initial expenditure level, the type and size of the school district, and the socio-economic class of the student example.

Levin listed three obstacles in obtaining schools that are more productive. These hurdles include (1) the lack of an information system; (2) seemingly limited mangement discretion over inputs; and (3) the absence of incentives to stimulate the achievement of the school's putative goals.[22]

Concerning the information system, educational systems need better information about their own operations and performance; priorities of the varied clientele need to be communicated better to decision makers; systems need better information about available educational technologies; and the systems need to develop alternative ways of making decisions.

20. Henry M. Levin, "The Effect of Different Levels of Expenditure on Educational Output," *Economic Factors Affecting the Financing of Education*, Chapter 6, pp. 186-188 (Gainesville, Florida: National Educational Finance Project, 1970).
21. *Ibid.*, p. 189.
22. *Ibid.*, pp. 194-197.

Management discretion is required at the individual classroom and individual student level as well as those levels more remote from the classroom. Also, it is only when the rewards of the organization are linked more closely to the goals of the schools that substantial improvements in dollar productivity will occur.

Parental involvement in the implementation of programs providing for equal educational opportunity is inherent in two decentralization models posed as solutions to today's educational problems—political decentralization and the market place. The benefits of the community school, which is a manifestation of political decentralization, are known generally and will not be treated in this chapter. However, decentralization via the market place will be considered briefly.

Educational Vouchers

Serious students of education and die-hard laissez-faire believers have generated enthusiasm for decentralization of education via the market place. Their fear is that pumping more money into public education will result in more of the same output at a higher cost. They have no faith in the burdensome bureaucracy of public education and reject the public education monopoly.

Adam Smith's idea for government to finance education by giving parents money to hire teachers is popular. A model is, of course, the G.I. Bill.

In December, 1969, the U. S. Office of Economic Opportunity (OEO) made a grant to the Center for the Study of Public Policy to support a detailed study of educational vouchers. Vouchers are a label for certificates which the government would issue to parents, parents would turn over the voucher to an eligible school, and the school would return the voucher to the government for cash. Models proposed are included in Table 3.

The Center stated that the nation must reallocate educational resources to expose "difficult" children to their full share of bright, talented, and sensitive teachers. Teachers prefer to teach those who are not difficult and who learn easily. To attract able teachers, the Center claimed that schools enrolling disadvantaged children must pay substantially more than schools which serve advantaged children. In addition, disadvantaged children must have more advantaged classmates since a student's classmates are probably his most important single resource.

The voucher system would be constructed upon another assumption, and that is that increasing parents' control over the kind of education their children receive will increase the chances that the children will get a good education.

Professor Coons of the University of California and his associates have proposed that Model 4—the effort voucher—be implemented. Schools would vary in their level of expenditure from the current level to that 2-3 times higher. Parent contributions would depend on the family's ability to pay and on the cost of the school the family chose. Perhaps 2 per cent of family income would be required for an $1,800 per pupil program, 1.5 per cent for a $1,200 program, 1.0 per cent for an $800 program, etc., and con-

TABLE 3
Seven Alternative Education Voucher Plans

1. *Unregulated Market Model*: The value of the voucher is the same for each child. Schools are permitted to charge whatever additional tuition the traffic will bear.
2. *Unregulated Compensatory Model*: The value of the voucher is higher for poor children. Schools are permitted to charge whatever additional tuition they wish.
3. *Compulsory Private Scholarship Model:* Schools may charge as much tuition as they like, provided they give scholarships to those children unable to pay full tuition. Eligibility and size of scholarships are determined by the EVA, which establishes a formula showing how much families with certain incomes can be charged.
4. *The Effort Voucher*: This model establishes several different possible levels of per pupil expenditure and allows a school to choose its own level. Parents who choose high-expenditure schools are then charged more tuition (or tax) than parents who choose low-expenditure schools. Tuition (or tax) is also related to income, in theory the "effort" demanded of a low-income family attending a high-expenditure school is the same as the "effort" demanded of a high-income family in the same school.
5. *"Egalitarian" Model*: The value of the voucher is the same for each child. No school is permitted to charge any additional tuition.
6. *Achievement Model*: The value of the voucher is based on the progress made by the child during the year.
7. *Regulated Compensatory Model*: Schools may not charge tuition beyond the value of the voucher. They may "earn" extra funds by accepting children from poor families or educationally disadvantaged children. (A variant of this model permits privately managed voucher schools to charge affluent families according to their ability to pay.)

Source: *Education Vouchers.* Cambridge, Massachusetts, Center for the Study of Public Policy, December 1970, p. 20.

sistent with Serrano, government would contribute the difference between what a family paid and what the school spent per pupil. A "truth in education" bill would have to be legislated so that parents might have valid information on the nature of each school. The Coons' model would result in the allocation of resources on the basis of willingess to pay rather than purely on ability to pay.

The fundamental political and pedagogic danger of the voucher plan is that the most popular schools would not be able to accommodate all students and would become, therefore, "over-applied." The over-applied schools would become privileged sanctuaries for students whom educators enjoy teaching. A necessary feature recommended by the Center was that the voucher system must provide economic incentives for enrolling so-called undesirable children.[23]

The concept of vouchers to increase both diversity and family power in education or to finance education by grants to parents represents an economic change that could improve the system of education or harm it severely. Legal restraints relative to racial segregation are clear but legal policy with respect to such issues as economic discrimination or religious involvement is less clear. A continuing problem, too, could be the control of "hucksterism" in the private sector of education. Grants to study the feasibility of the voucher system have been made to three West coast communities and only after controlled experiments of this kind have been conducted can valid decisions be made.

MICRO SYSTEMS

Local Resource Management

Previous material in this chapter was a description of the macro systems (national and state) of which the local school community is a subsystem. The complexity of the interrelationships between the public and private sectors, school finance and public finance, local-state-federal governments, public education and the GNP, etc., indicate the constraints under which local school communities operate plus the opportunities to gain access to state and national wealth.

Within the limits of the sets of constraints and opportunities, the local school district must allocate scarce resources to meet the unlimited needs of its clients. System analysis is an approach to decision making that emphasizes the following: (1) definition of educational problems; (2) development of alternative programs; (3) analysis of alternative solutions; and (4) recommendation of preferred programs.

A management model can be designed as sets of flows of information and activities of participants. A generic management model is outlined in

---

23. *Education Vouchers* (Cambridge, Massachusetts: Center for the Study of Public Policy, December 1970), p. 348.

Figure 1. Implementation of the model should result in a school organization that generates new ideas, is responsive and adaptive.

The simplest type of systems model of an organization relates inputs to outputs. Inputs are the material and human resources which are converted in the system to outputs. They include such factors as teachers, students, equipment, and dollars. The outputs will be levels of developed human resources—i.e., people with problem-solving ability, social and political skills, and marketable vocational capabilities. Management is the linking mechanism between the inputs and outputs in the conversion process.

A mechanism for evaluating the system's output in relation to its objectives must exist as a means of regulating the inputs and operations in the system. This is done by a feedback mechanism whereby a system of such complexity may be monitored in order to remain sensitive and adaptive to its internal and external environments. The system model now consists of input-function-output feedback. Communication must occur not only between system outputs as related to inputs and functions, but also within and between each function level described in Figure 2.

Critical educational needs for school districts include a continuous self and organizational renewal via participative and/or decentralized management systems and structures via system concepts and improved problem finding aids. An excellent instrument for development of a self-renewal system of the kind just described is the local school district budget. It is the most concrete planning evidence, and if the budgetary process is ideal it involves other administrative processes beyond planning, including organizing, communicating, evaluating, and re-appraising.

The Budget—A Multipurpose Tool

The local school district budget can be used to perform at least four major functions: (1) superintendent's request of the board of education for allocation of resources; (2) report to the citizens; (3) controlling; and (4) planning. It is a school system's plan of operations that is expressed in words, numbers and dollars. It expresses the conclusion reached by the system in pursuance of matters that have been studied, planned, and programmed in one segment of time.

Sophisticated school systems generally develop several types of budgets including: (1) a current budget (pertains to operations in a fiscal year); (2) a long-term budget (a plan comprising a community's educational aspirations for a future period of five to ten years); (3) a capital budget (capital outlay plans for several years); (4) a building budget

Figure 1. A Generic Management Model

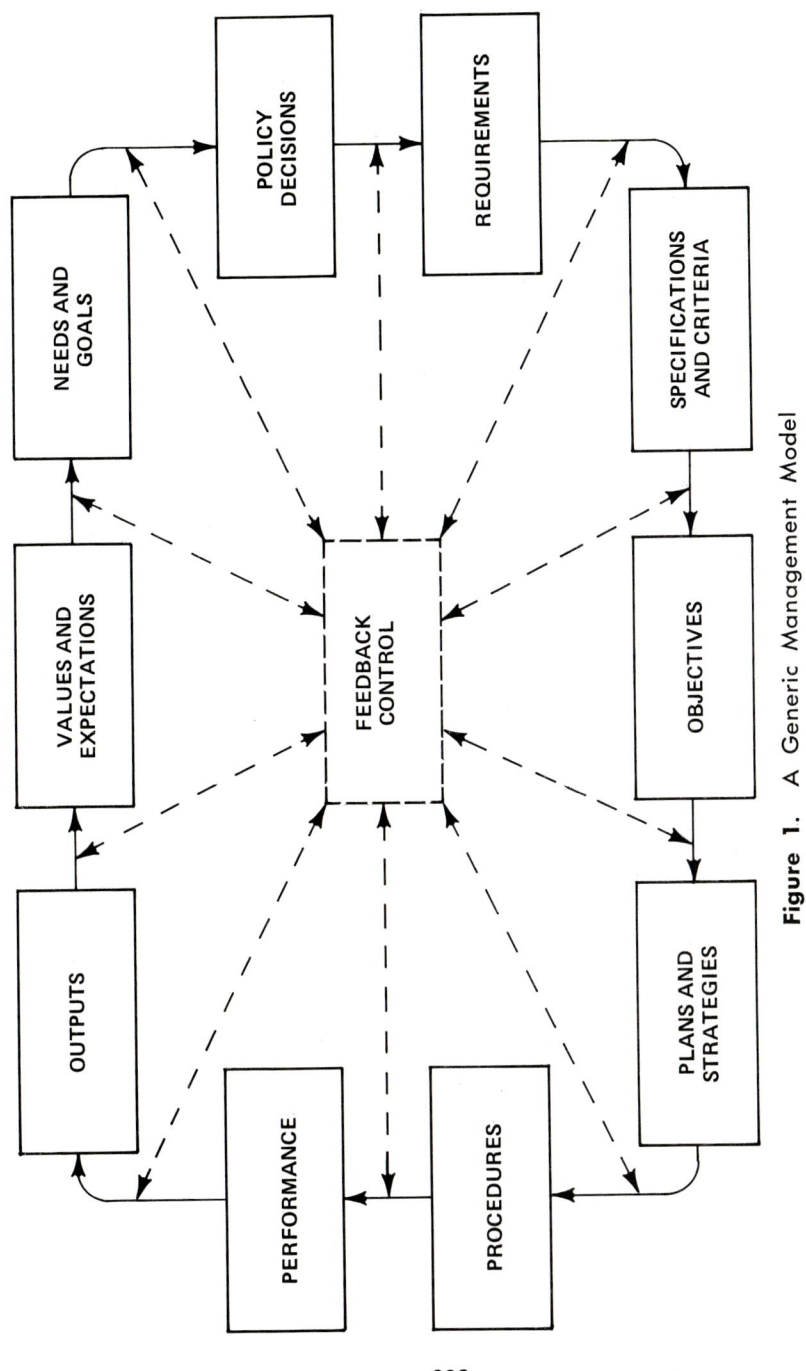

Source: Donald R. Miller, An Overview of a Process for Managing Change, San Mateo, California: OPERATION PEP, April 1970.

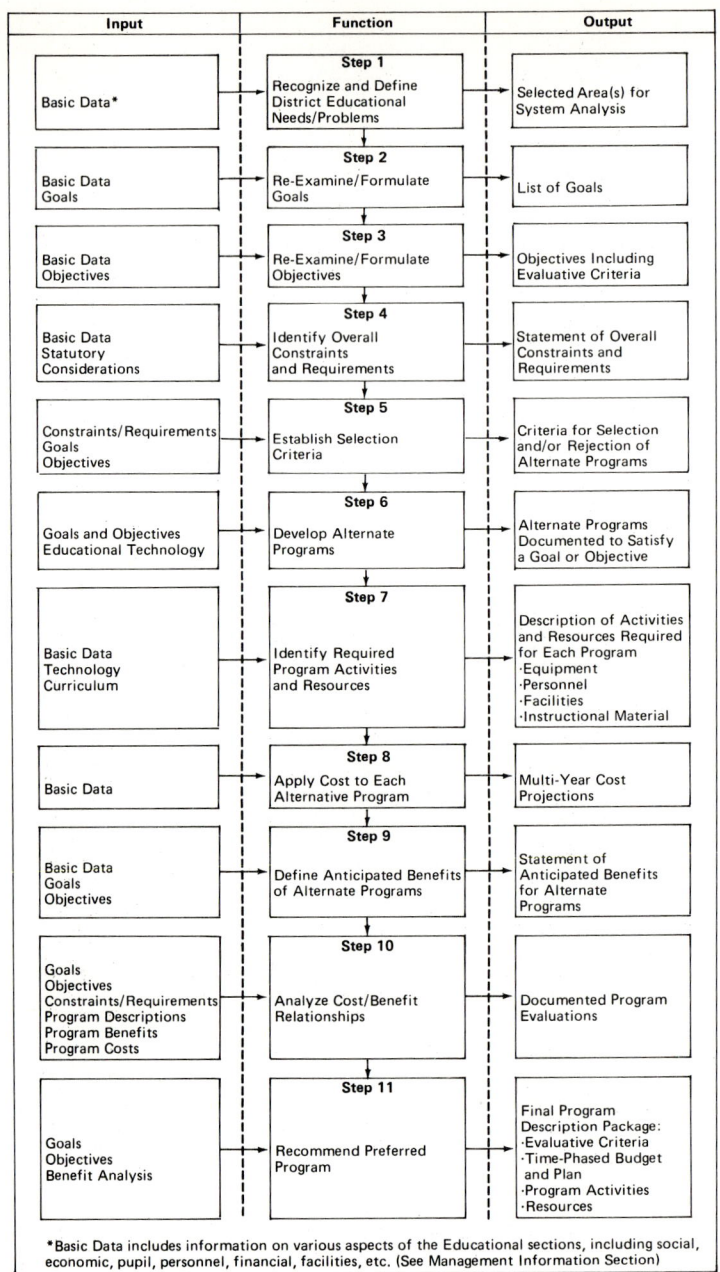

Input	Function	Output
Basic Data*	**Step 1** Recognize and Define District Educational Needs/Problems	Selected Area(s) for System Analysis
Basic Data Goals	**Step 2** Re-Examine/Formulate Goals	List of Goals
Basic Data Objectives	**Step 3** Re-Examine/Formulate Objectives	Objectives Including Evaluative Criteria
Basic Data Statutory Considerations	**Step 4** Identify Overall Constraints and Requirements	Statement of Overall Constraints and Requirements
Constraints/Requirements Goals Objectives	**Step 5** Establish Selection Criteria	Criteria for Selection and/or Rejection of Alternate Programs
Goals and Objectives Educational Technology	**Step 6** Develop Alternate Programs	Alternate Programs Documented to Satisfy a Goal or Objective
Basic Data Technology Curriculum	**Step 7** Identify Required Program Activities and Resources	Description of Activities and Resources Required for Each Program ·Equipment ·Personnel ·Facilities ·Instructional Material
Basic Data	**Step 8** Apply Cost to Each Alternative Program	Multi-Year Cost Projections
Basic Data Goals Objectives	**Step 9** Define Anticipated Benefits of Alternate Programs	Statement of Anticipated Benefits for Alternate Programs
Goals Objectives Constraints/Requirements Program Descriptions Program Benefits Program Costs	**Step 10** Analyze Cost/Benefit Relationships	Documented Program Evaluations
Goals Objectives Benefit Analysis	**Step 11** Recommend Preferred Program	Final Program Description Package: ·Evaluative Criteria ·Time-Phased Budget and Plan ·Program Activities ·Resources

*Basic Data includes information on various aspects of the Educational sections, including social, economic, pupil, personnel, financial, facilities, etc. (See Management Information Section)

Figure 2. System Analysis Process

Source: *Conceptual Design for a Planning, Programming, Budgeting System,* Sacramento: California State Department of Education, 1969, p. 19.

(revenues and expenditures of a building project); and (5) a special project budget (projects omitted or separated from current expenditures).[24]

Ideally, the preparation of the current or operating budget is a year-round job. As an example of budget preparation activities of a major school system, Figure 3 is Milwaukee Public Schools' PERT presentation as prepared by the district's Divison of Planning and Long-Range Planning. The activities are listed on the two pages following Figure 3. It should be noted that Milwaukee's fiscal year is a calendar year.

In their original development, both PERT and CPM were essentially time-oriented.[25] Managers used them as planning tools to estimate the time required to complete a proposed project. Thus, time schedules for project activities (as exemplified in Figure 3) could be established and control devices to check schedule times against actual times for activity durations or event occurrences were effected.

PERT and CPM cost systems have been developed and are simple but importantly different from that of most cost accounting systems. The difference in essence is that costs are to be measured and controlled primarily on a project basis rather than according to the functional organization of an institution. Micro cost centers are formed around individual activities and the rationale is that responsibility for expenditures should coincide with responsibility for managing that which gives rise to the expenditures. In education, this would mean that teachers would be given control of resources for the activities they were supervising. Some believe that it is this kind of decentralization in fiscal management that can lead to the desired revolution in public education.

PPBS

The Planning-Programming-Budgeting System (PPBS) is an integrated management system that places emphasis on the use of analysis for program decision making. The purpose of PPBS is to provide management with a better analytical basis for making program decisions and for putting such decisions into operation through the systematic integration of the planning, programming, and budgeting functions. The PPBS aspects and tasks involved include:[26]

---

24. Garvue, *Modern Public School Finance*, pp. 112-113.
25. PERT and CPM are acronyms for "program evaluation and review technique" and "critical path method."
26. John A. Evans, *PPBS: From Budgeting and Programming to a Planning, Programming, Budgeting System* (San Mateo, California: OPERATION PEP, April 1970).

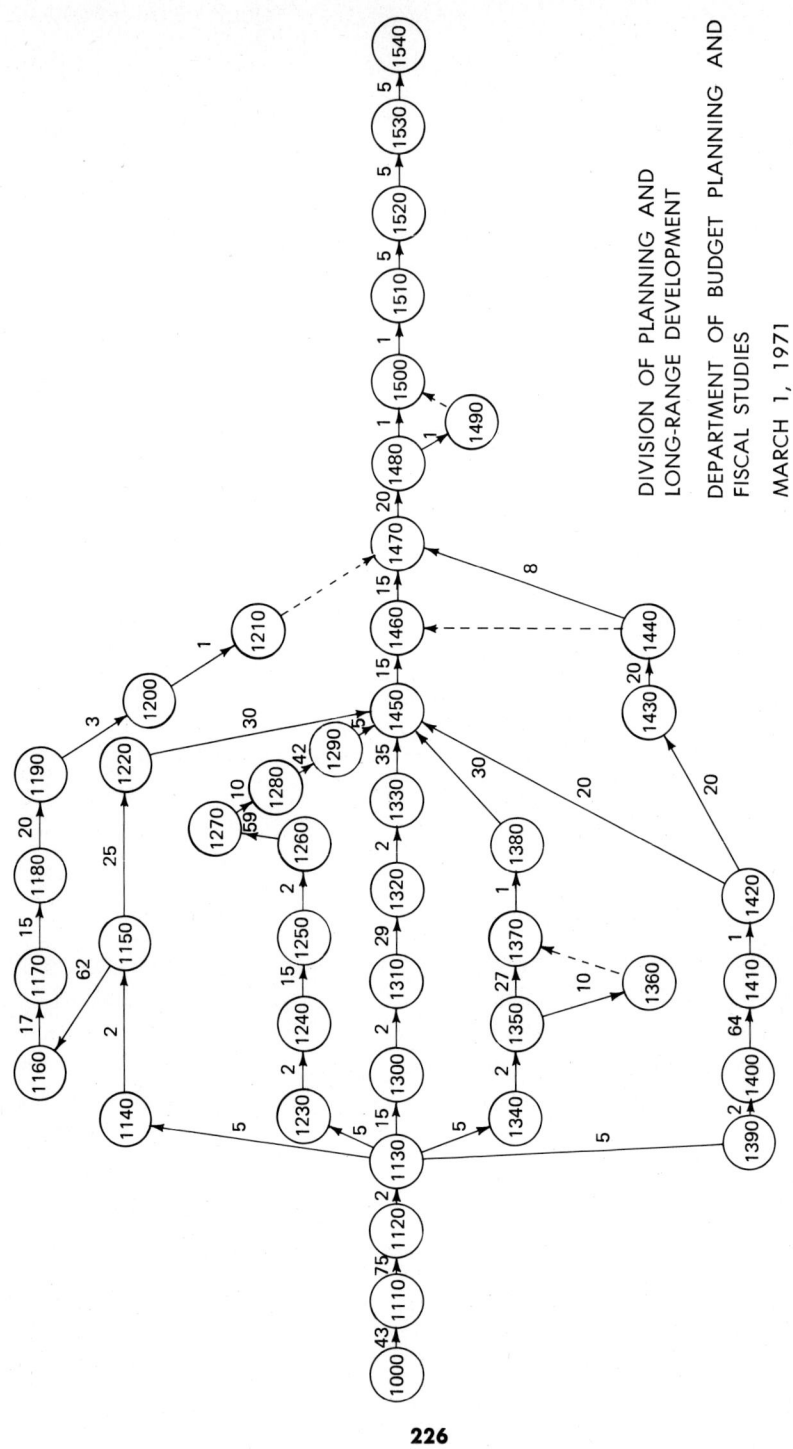

Figure 3. Milwaukee Public Schools 1972 Budget Preparation Network

1972 BUDGET PREPARATION ACTIVITIES

Activity	Description	Limiting Dates	Days Duration
1000-1110	System-wide Thrusts	7/1-8/27	43.0
1110-1120	School Planning Sessions	9/1-12/18	75.0
1120-1130	Identify Budget Document Objectives	11/1-12/1	2.0
1130-1140	Prepare Program Improvement Proposal Forms	12/1-1/4	5.0
1130-1230	Prepare Plant Improvement Forms	12/1-12/18	5.0
1130-1300	Prepare School Instruction Budget Forms	1/15-3/5	15.0
1130-1340	Prepare Central Office Instruction Budget Forms	1/15-2/12	5.0
1130-1390	Prepare Central Office Operations Budget Forms	1/15-2/12	5.0
1140-1150	Send Program Improvement Proposal Forms	1/4-1/5	2.0
1150-1160	Prepare New Program Improvement Proposals	1/6-3/31	62.0
1150-1220	Prepare Continuing Program Improvement Projects	1/6-4/23	25.0
1160-1170	Review New Program Improvement Proposals	4/1-4/30	17.0
1170-1180	New Program Improvement Proposal Final Revision	5/3-5/24	15.0
1180-1190	Appointment-Instruction Committee Recommendations	5/25-6/25	20.0
1190-1200	Finance Committee Recommendations	5/26-6/28	3.0
1200-1210	School Board Authorization	6/29	1.0
1210-1470	Dummy	....	....
1220-1450	Analyze Continuing Program Improvement Projects	4/23-6/11	30.0
1230-1240	Send Plant Improvement Forms to Schools	12/19-12/20	2.0
1240-1250	Schools Prepare Plant Improvement Requests	12/21-1/21	15.0
1250-1260	Schools Submit Plant Improvement Requests	1/22-1/23	2.0
1260-1270	Analyze Plant Improvement Requests	1/24-4/22	59.0
1270-1280	Repair Review Board Preliminary Decisions	3/3-4/23	10.0
1280-1290	Review Plant Improvement Decisions with Schools	4/26-6/11	42.0
1290-1450	Repair Review Board Final Decisions	6/9-6/15	5.0
1300-1310	Send Instruction Budget Forms to Schools	3/8-3/9	2.0
1310-1320	Schools Prepare Instruction Budgets	3/10-4/22	29.0
1320-1330	Schools Submit Instruction Budgets	4/23-4/24	2.0
1330-1450	Analyze School Instruction Budget Requests	4/26-6/11	35.0
1340-1350	Send Instruction Program Budget Forms to Central Office Coordinators	3/8-3/9	2.0
1350-1360	Program Coordinators Consult with Schools	3/10-4/22	10.0
1350-1370	Program Coordinators Prepare Instruction Budgets	3/10-4/22	27.0
1360-1370	Dummy	....	....
1370-1380	Program Coordinators Submit Instruction Budgets	4/23	1.0
1380-1450	Analyze Central Office Instruction Program Budgets	4/26-6/11	30.0
1390-1400	Send Central Office Operations Budget Forms to Division Heads	3/15-3/16	2.0
1400-1410	Division Heads Prepare Central Office Budgets	3/17-6/17	64.0
1410-1420	Submit Central Office Budgets	6/18	1.0
1420-1430	Analyze Central Office Business Budgets	6/21-7/16	20.0
1420-1450	Analyze Central Office Education Budgets	6/21-7/16	20.0
1430-1440	Secretary-Business Manager Budget Decisions	7/19-8/6	20.0
1440-1460	Dummy	....	....
1440-1470	Secretary-Business Manager Revenue Estimate	9/1-9/10	8.0
1450-1460	Superintendent Council Budget Decisions	7/19-8/6	15.0
1460-1470	Board Building-Finance Committee Public Work Sessions	8/9-8/27	15.0
1470-1480	Prepare Budget Document	8/9-9/24	20.0
1480-1490	Board Building Committee Recommendations	9/29	1.0
1480-1500	Board Finance Committee Recommendations	9/30	1.0
1490-1500	Dummy	....	....
1500-1510	Building-Finance Committee Recommendations to Board	10/1	1.0
1510-1520	Prepare Budget Summary	10/4-10/8	5.0
1520-1530	Board Hearings on Budget	10/11-10/15	5.0
1530-1540	Board Final Budget Decisions	10/18-10/20	5.0

228   FINANCING THE EDUCATIONAL PROGRAM

Aspects	Tasks
Structural Aspect Goal/Objective Hierarchies	Setting Objectives Assigning Activities to Programs
Analytical Aspect Cost/Effectiveness Studies	Determining Resource Requirements Costing Programs Developing Criteria Identifying Alternatives Determining Cost Effectiveness of Alternatives
Information System Aspect Progress and Analysis	Evaluating Alternatives Updating Program

Currently, a number of school districts throughout the nation are taking the first steps to develop PPBS. Referring to the aspects and tasks just outlined, the weakest are the tasks of developing criteria, identifying alternatives and determining cost effectiveness of alternatives. The structural aspects and information system aspects are shaping into reasonably sophisticated form. In other words, cost accounting has been improved, data are becoming available relative to the comparative costs of instructional programs (elementary v. secondary, art v. English, etc.) but the analytical work is virtually nonexistent.

As an example, however, of an accomplishment in the analytical, the Clark County School District (Las Vegas, Nevada) developed a pilot PPBS budget for the district's Business and Finance Division. Programs within the divisional budget included:

    Associate Superintendent of Business and Finance
    Accounting Department
    Data Processing Department
    Financial Services Department
    Food Services Department
    Purchasing, Warehousing & Inventory Management Department
    Supplies and Equipment Department
    Transportation Department

Each department then included its program, objectives and resource allocation. As an example, the format for the Accounting Department was as follows:

    Department—Accounting

    Program   —Accounts Payable

    Broad Objective—Pay non-salary obligations in compliance with district regulations and legal requirements

Specific Objective—Pay 90 per cent of all bills within 30 days after the receipt of goods or services; combine our 1970-71 vendor files with those of the Purchasing Department for a more efficient operation.

Services—Maintain files to document payments
    a. Paid warrants and invoices
    b. Completed purchase orders and requisitions
Maintain record of utility expenditures by location and advise proper persons of unusual differences
Take credits and discounts
Provide replacement warrants where necessary
Check vendor statements

Evaluation—After each payment cycle (regular board meeting)
10 per cent of all payments made are audited to determine if objectives are being met; a review of what was accomplished in combining vendor files by June 30, 1972

Resources (1971-72)	% of time	No. of persons	Cost	(1972-73) % of Time	No. of Persons	Cost
Director	10	1	$ 2,034	10	1	$ 2,034
Senior Accountant	25	1	3,980	25	1	3,980
Account Clerks	100	5	38,850	100	5	39,503
Secretary	15	1	1,251	15	1	1,313
Supplies & Expense			1,800			1,890
		8	$47,915		8	$48,720

Costs were also projected into 1973-74. Relative to the analytical aspect, an exhibit form for the purpose of determining the time lapse between receipt of materials or services and the mailing of checks is Figure 4.

Analytical Developments

Some of the most sophisticated analysis is being completed by the Education Turnkey Systems. The private corporation had the responsibility for determining the costs involved with each program in the Office of Economic Opportunity experiments in performance contracting. Verbatim material from the Final Report to the Office of Economic Opportunity follows from Analysis Summary through Special Sensitivity and Trade-Off Analyses.[27]

ANALYSIS SUMMARY. The objective of the Analysis Summary is to provide the administrator with a general description of the economics of the instructional system being analyzed. The summary can provide that information which he

---

27. *Performance Incentive Remedial Education Experiment* (Washington, D.C.: Education Turnkey Systems, Final Report to OEO, BOO-5114, (31 August 1971), pp. 157-159.

230 FINANCING THE EDUCATIONAL PROGRAM

To: Director of Accounting   Date: _____
From: Auditor
Subject: An analysis of the _____ (Board Date) A/P to determine the time lapse between receipt of materials or services and mailing of checks.

_____ % of all CC-4's, CC-34B's, and CC-34A's were examined.

_____ aggregate days ÷ _____ items = overall days.

CC-4 aggregate days _____ ÷ _____ items = _____ days average

CC-34B aggregate days _____ ÷ _____ items = _____ days average

CC-34A aggregate days _____ ÷ _____ items = _____ days average

Items paid in   Total items
30 days or less   examined

1. _____ ÷ _____ = _____ % of CC-4's (travel claims) were paid in 30 days or less.

2. _____ ÷ _____ = _____ % of CC-34B's (special purchase authorizations) were paid in 30 days or less.

3. _____ ÷ _____ = _____ % of CC-34A's (purchase orders) were paid in 30 days or less.

Lines 1+2+3   Lines 1+2+3

4. _____ ÷ _____ = _____ % of CC-34B's and CC-34A's examined have been paid within 30 days from receipt of materials or services to the date checks are scheduled to be mailed.

Comments: _____

_____

_____

Figure 4. Clark County School District
Las Vegas, Nevada
EXHIBIT 2

feels would be appropriate in light of his particular situation regarding budget projections, his interest in the effectiveness of the program, and his willingness to initiate major management changes. The Analysis Summary gives the existing costs per unit of achievement, and the total costs for the program cycle, and could include such factors as drop-out rates and repeater rates. It also indicates the opportunity costs associated with certain managerial decisions. For example, the opportunity cost of a facility which is only used 20% of the time is extremely high since interest is being paid on the basis of a 100% utilization rate. The Analysis Summary would indicate to the administrator the costs of not utilizing the facility to a greater extent.

ECONOMIC STRUCTURE ANALYSIS. The purpose of the structure analysis report is to indicate in great detail the way and amount by which each of several hundred economic factors contributes to the total costs of the instructional program. In order to portray this breakdown of total costs, the factors are first broken down by function (e.g., classroom instruction) then by resource type utilized in that function (e.g., teacher time, books, audio-visual equipment) and then by particular characteristics of each resource in providing that particular function (e.g., teacher-student ratio, audio-visual equipment utilization rates). This particular report can be used as an information source or as a basis for a program modification. As an information source, the report can be used by the administrator to approximate the impact of alternative decisions on the total cost of instruction. If the administrator desires to use the report for program modification, he can trace backwards from the total cost to the costs of each function and then to the resources types which consume costs. From this preliminary analysis, he can determine the direction for program improvement based solely on the economics of the existing instructional program. He may also want to use the Model as a basis of reviewing proposed program modifications and radically different methods of instruction in terms of the way their projected costs compare with the economics of the existing program.

SENSITIVITY ANALYSES. The purpose of the Sensitivity Analysis report is to indicate how sensitive the toal cost of operations is to changes in each of the economic factors or program design characteristics. For example, the administrator may want to determine the relative economic importance of certain factors which, based on his experience and knowledge, he feels contribute to educational effectiveness, (e.g., books, audio-visual equipment, etc.) compared with that of other factors which he may not believe contribute as much to student performance increases (e.g., physical education equipment, classroom size, student-teacher ratio, etc.). The Model allows the economic factors to be described in terms of various levels of total program costs. For example, the impact on total cost of 5, 10, and 25 per cent increases and decreases in the annual pay of the teacher is described in a way that graphically displays the sensitivity of total cost to changes in that factor. In short, the report is meant to be used in conjunction with the Economic Structure Analysis discussed above as a basis of further program modification and improvement. It provides cost data in a manner which complements the administrator's judgment about the educational effectiveness of changes in the resources consumed in the program.

ECONOMIC FACTOR RANKING. The purpose of the Economic Factor Ranking report is to display, in order, those economic factors which have the greatest impact on total costs within the school system. If the administrator is faced with a budget cut or, within existing budget levels, with a demand for increases in certain portions of the budget (e.g., teachers' salaries increases), he will know what economic factors he should consider in terms of priorities for making marginal percentage cuts which will show large absolute dollar savings. For example, the Model would indicate the increase in average teacher salary which would be made possible if the entire budget were increased by 1%. It would also indicate various economic factors which would approximate an equal cost trade-off. An increase in teachers' salaries resulting in a total budget increase of 1% could be absorbed by a decrease in expenditures on books or by an increase in student-teacher ratio. Since this report ranks the program characteristics according to the degree to which each contributes to total costs, the administrator may identify, using his judgment as to the learning value of each factor those characteristics which appear costly relative to their effectiveness.

COMPARATIVE SUMMARIES. After conducting some type of pilot or experimental program in his district, the administrator might be interested in the turnkeying of such a program into the normal structure of the district. This report, which displays selectable characteristics of as many as six alternate instructional programs, allows him to readily compare his current district program with the pilot program (or any other programs for which he at least has data available).

To assist the administrator in a turnkey decision, the Economic Structure Analysis would first be examined for each program being compared. Those factors which are most amenable to change as well as those which would be most difficult, in a political sense, to change can be identified in the Structure Analysis.

The administrator would then have a Comparative Summaries report run for those factors which would be difficult to change, and probably another report run for those factors more amenable to change. These two reports would display, on a program-by-program comparison basis, which program relied most heavily on changes in cast-in stone factors. Such a program would be a less likely candidate than another which involved more manageable factors.

This report could be viewed as a turnkey shopping list showing all at the same time the important factors involved in a number of alternative programs as well as the requirement each of these programs has for each of the factors. Finally, to add relevance to all this information, the administrator can also request the cost per unit of achievement for each of the programs being compared.

SPECIAL SENSITIVITY AND TRADE-OFF ANALYSES. When the administrator has determined the economic factors which he wants to consider for program modification, improvement or evaluation, he can then request special reports which explore the relationship between these factors and other characteristics of the program. For example, the Special Sensitivity Analysis is designed to show how changes in any one economic factor (e.g., student-teacher ratio) affect the requirement for any other economic factor (e.g., number of staff required). The

Trade-Off Analysis is designed to show equal-cost alternatives for funds allocated between two specific factors. This report answers the following type of question: Given a fixed budget constraint and an increase in teachers' salaries of 5%, how much of an increase in student-teacher ratio would be required to meet the fixed budget? These analyses, in short, assist the administrator in allocating funds so as to maintain a fixed total cost, as well as to dramatize the impact that change in any one factor or program design feature would have on limiting resources for any other factor.

COST EFFECTIVENESS. Cost-effectiveness analysis is the process by which costs and certain benefits associated with program outputs are related and studied by decision makers in the determination of priorities and the allocation of resources. Three formulas have been suggested:[28]

$$\text{Efficiency Ratio} = \frac{\text{Standard Cost/Unit of Output}}{\text{Actual Cost/Unit of Output}}$$

$$\text{Time Effectiveness Ratio} = \frac{\text{Actual Output/Time Period}}{\text{Planned Output/Time Period}}$$

$$\text{Educational Productivity: } \frac{\text{Output}}{\text{Input}} = \frac{\text{OEd}}{\text{IEd}}$$

An example of the latter would be: OEd, a ratio of semester hours of college credit earned by advanced placement in mathematics, expressed in per cent of total semester hours earned; and IEd, a ratio of average teaching costs in mathematics, expressed in per cent of total amount expended.

Cost effectiveness is considered to be a tool for micro-analysis while cost-benefit and cost-utility analyses are termed intermediate- and macro-level, respectively. The purpose of cost-benefit study is to determine the alternative that yields the greatest benefits for any given cost relative to the achievement of an objective. In cost-utility research it is assumed that for any particular program or activity the cost can be assigned a number and the utility (or benefit) of the program can be assigned a number. The ratio of the two numbers, the cost-utility ratio, can then be compared with the ratio for any other project.

An outcome of the movement toward effecting cost effectiveness studies is the development of cost and educational output accounting systems. A weakness in costing procedures in the past has been to ignore such significant indirect costs such as the depreciation and obsolescence of buildings as well as the imputed interest on invested capital and the stu-

---

28. Richard H. P. Kraft, *Cost Effectiveness: Analysis of Vocational-Technical Education Programs* (Tallahassee, Florida: Department of Educational Administration, 1969), p. 44.

dents' foregoing earnings. Models of costing are becoming an important element in the current literature on education as a production function.[29]

Educational Planning

Three approaches to educational planning and to educational policy development are: (1) the social demand approach; (2) the rate-of-return approach; and (3) the manpower-requirements approach.

The social demand approach traditionally has been the one relied upon—i.e., places in school are provided for everyone who wants to go to school. A liability of such a framework is that when there are insufficient resources to meet all the demands, there is no rational way to determine priorities.

Ranking educational programs according to their rates of return is exemplified by the work of Hansen when he found that the social internal rates of return to three levels of education in the United States were: primary education, 14.0 per cent; secondary education, 11.4 per cent; and higher education, 10.2 per cent.[30] Thus, primary or primary and secondary education would be increased relative to higher education. A major criticism of this approach is that it is based on a concept dealing with averages—the key question is not what the average man earned but what the last man earned and there is no way of determining who the marginal man is. A second criticism is that past rates of return are inappropriate for planning in that future rates should be used.[31]

A major defect in the manpower-requirement approach is the assumption that the educational requirements for each job are rigid when in fact a range of educational backgrounds is typically seen for almost every job, including that of school administration. Manpower planning ignores the cost of education and neglects low-skill labor needs.

For the future, an eclectic approach seems to be the most feasible. Adding the rate-of-return and manpower-requirement approaches to the social demand method at the local district level will force the district to expand its data base and to ask itself harder questions about programs and their relevance.

---

29. J. Alan Thomas, *The Productive School* (New York: John Wiley & Sons, Inc., 1971), pp. 31-55.

30. W. Lee Hansen, "Total and Private Rates of Return to Investment in Schooling," *Journal of Political Economy*, 71 (1963): 128-40.

31. Daniel C. Rogers and Hirsch S. Ruchlin, *Economics and Education* (New York: The Macmillan Company, 1971), p. 225.

CONCLUSIONS

The crisis in American education has been building a long time. What has been termed the "iceberg" image—nine-tenths of the troubles have been submerged beneath a sea of public complacency and preoccupation with other matters—is being destroyed. Problems are pressing in from every side and there is a clear demand for action that will enhance individual student learning, the effectiveness of schools and the quality of life nationally.

Rapid population growth and mobility, country-to-city migration, unpredictable economic and social changes brought by technology and the disproportionate claims of the military on the gross national product have made schools the victims of many conditions beyond their control. However, numerous problems are internal in that systematic approaches to instruction and management have been lacking. Lack of a concentrated research, development, and application network, too, makes the educational transformation more difficult.

Research expenditures amount to no more than one-fourth of 1 per cent of the nation's expenditures for schools, colleges, universities, and other educational enterprises—that is, a total of no more than $125 million annually. This represents progress, however, since a few years ago the ratio was only one-tenth of 1 per cent.[32] In industry, about 4 per cent of net sales is spent on basic and applied research ($18 billion annually), while in health, an amount equivalent to nearly 5 per cent of expenditures ($2.5 billion) is dedicated to research. Those who believe that education needs to be more capital intensive may be shocked to learn that more dollars are spent in one week for programming three major television networks than in a year for all educational television.[33]

A well-designed plan without sufficient funds and without sufficient incentives to capture the support of operational level functionaries will bring little positive change. Elementary and secondary education needs development of incentive systems desperately, particularly in districts with a disproportionate number of children who have been categorized as difficult learners.

By now it should be obvious that reliance upon system analysis or the "new cult of efficiency" alone will not bring about the transformation of American education. The state of the analytical art is relatively crude, although the innovators offer exciting possibilities. PPBS today is basically

---

32. *To Improve Learning* (Washington, D.C.: Committee on Education and Labor, House of Representatives, March 1970), p. 63.

33. *Ibid.*, p. 23.

a cost accounting system which obviously is an improvement over traditional fiscal accounting. However, decison making is as complicated as ever, the reason being that decisions are value-laden and our knowledge and use of the social sciences hasn't given us the capability to resolve basic value issues. *Serrano v. Priest* is a breakthrough on a major issue involving redistribution of income via increased services. The redistribution of educational services and resources, and a resultant equity in the context of a national community is an important first step in the educational transformation. An implementation of a national Right to Equity program may be the incentive most essential.

Selected References

Becker, Gary S. "Investment in Human Capital: A Theoretical Analysis." *Journal of Political Economy* (Supplement on Investment in Human Beings), Vol. 70, No. 5, Part 2 (1962), pp. 9-49.

Benson, Charles S. *The Cheerful Prospect.* Boston: Houghton Mifflin, 1965.

———. *The Economics of Public Education.* Boston: Houghton Mifflin, 1968.

Blaug, Mark. "Approaches to Educational Planning." *Economic Journal* 77 (1967): 262-87.

Boulding, Kenneth E. "Economics as a Moral Science." *American Economic Review* 59 (1969): 1-12.

Bowles, Samuel. "The Efficient Allocation of Resources in Education." *Quarterly Journal of Economics* 81 (1967): 189-219.

Bowles, Samuel, and Levin, Henry. "The Determinants of Scholastic Achievement." *Journal of Human Resources* 3 (1968): 3-24.

Bowman, Mary Jean. "Schultz, Denison and the Contribution of 'Eds' to National Income Growth." *Journal of Political Economy* 72 (1964): 450-64.

Cain, Glen G., and Stromsdorfer, Ernst W. "An Economic Evaluation of Government Retraining Programs in West Virginia." In *Retraining the Unemployed*, edited by Gerald G. Somers. Madison: University of Wisconsin Press, 1968.

Coons, John E. et al. *Private Wealth and Public Education.* Cambridge, Massachusetts: The Belknap Press, 1970.

Denison, Edward F. *Why Growth Rates Differ.* Washington, D.C.: The Brookings Institution, 1967.

Garms, Walter A. "A Multivariate Analysis of the Correlates of Educational Efforts of Nations." *Comparative Education Review* 12 (1968): 281-99.

Garvue, Robert J. *Modern Public School Finance.* London: Macmillan Company, 1969.

Gerwin, Donald. *Budgeting Public Funds.* Madison: The University of Wisconsin Press, 1969.

Hack, Walter, and Woodard, Francis O. *Economic Dimensions of Public School Finance*. New York: McGraw-Hill Book Company, 1971.

Hanoch, Giora. "An Economic Analysis of Earnings and Schooling." *Journal of Human Resources* 2 (1967): 310-29.

Herber, Bernard P. *Modern Public Finance: The Study of Public Sector Economics*. Homewood, Illinois: Richard D. Irwin, Inc., 1971.

Hettich, Walter. "Equalization Grants, Minimum Standards, and Unit Cost Differences in Education." *Yale Economic Essays* 8 (1968): 3-55.

Hirschleifer, J. "On the Theory of Optimal Investment Decision." *Journal of Political Economy* 66 (1958): 329-52.

James, H. Thomas: *The New Cult of Efficiency*. Pittsburgh: University of Pittsburgh Press, 1969.

Johns, Roe L., and Morphet, Edgar L. *The Economics and Financing of Education*. Englewood Cliffs, New Jersey: Prentice-Hall, 1969.

Johns, Roe L. et al., eds. *Economic Factors Affecting the Financing of Education*. Gainesville, Florida: National Educational Finance Project, Vol. 2, 1970. Also, *Planning to Finance Education,* Vol. 3, 1971; *Status and Impact of Educational Finance Programs,* Vol. 4, 1971; and *Alternative Programs for Financing Education*, Vol. 5, 1971.

Katzman, Martin. "Distribution and Production in a Big City Elementary School System." *Yale Economic Essays* 8 (1968): 201-56.

Mincer, Jacob. "On the Job Training: Costs, Returns, and Some Implications." *Journal of Political Economy*, Vol. 70, No. 5, Part 2 (1962), pp. 50-79.

Morton, Anton S. "Supply and Demand of Teachers in California." *Socio-Economic Planning Science* 2 (1969): 487-501.

Rogers, Daniel C., and Ruchlin, Hirsch S. *Economics and Education*. New York: The Free Press, 1971.

Samuelson, Paul A. *Economics: An Introductory Analysis*. New York: McGraw-Hill, 1970.

Spiegelman, Robert G. et al. *A Benefit/Cost Model to Evaluate Educational Programs*. Menlo Park, California: Stanford Research Institute, 1968.

Weidenbaum, Murray L. *The Modern Public Sector*. New York: Basic Books, Inc., 1969.

Wiest, Jerome D. *A Management Guide to PERT/CPM*. Englewood Cliffs, New Jersey: Prentice-Hall, Inc., 1969.

Wiseman, Jack. "Cost-Benefit Analysis in Education." *Southern Economic Journal* 32 (1965): 1-12.

9

JOHN H. JOHANSEN

Serving the Client System

SOCIAL SYSTEMS

School districts can be viewed in many ways. They can be described as political entities or geographical areas. They can also be looked upon in terms of their purpose, that of providing educational services. In this chapter, they are considered to be all of the aforementioned, but primarily they will be viewed and examined as social systems. In a most general definition, a social system is a group of individuals organized to accomplish a purpose. It may be a highly structured formal bureaucracy accomplishing many purposes and serving extremely diverse and pluralistic groups, or it may be very informal and small. A state or federal government can be thought of as a social system, as can a four couple contract bridge group or a few friends meeting regularly in a local coffee shop.

A school district as a social system is composed of individuals and groups, pluralistic in nature, yet similar in that each individual and group is in some way or another a part of a public entity whose overall goal is to provide educational services. As such, it is part of a supra-system and has sub-systems. Figure 1 illustrates a classification of social systems showing a school district as a sub-system of other supra-systems.

In Figure 1, and for purposes of analysis, artificial boundaries are drawn. The State and Nation are shown as being supra to the school district. They are considered to be in the external environment of the school district. The fact that they interact with one another is illustrated in that their boundaries touch each other at a common point. The arrows indicate information flow from one social system to another. An illustration of information flow from the national supra-system to the local school district

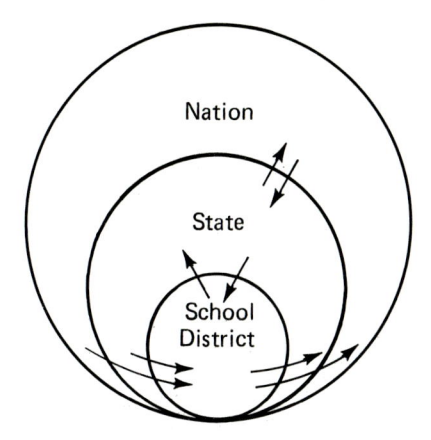

Figure 1. Supra- and Sub-systems

could be a court order to desegregate their school system. An illustration of information flow from the state could be a requested change in state funding procedures. By the same token, information flow from the local system to the national level could be their responses to the court order and the requested state funding change. The nature of the response to received information will vary, perhaps from no response to a violent reaction. Nevertheless, the "status quo" of the social system will be disturbed by information and a reaction to the disturbance is to be expected. In reality, the exchange of information occurs across the artificial boundaries in many places, through a variety of media, and with great rapidity. Multiple events occuring simultaneously within each system, and in all systems, interact in complex ways generating other events and more information exchanges. Simple cause and effect relationships are inadequate to explain the phenomena.

It is important to note also that the school district has many possible sub-systems. Each individual or group of individuals could be categorized as a sub-system. They could be thought of in terms of occupational groups such as farmers, businessmen, professionals, blue collar workers, or retired persons; or in terms of economic status, i.e., wealthy or poor; or in terms of ethnicity, black, white, Spanish-American, or Indian; or by religious designation, Protestant, Catholic, or Hebrew; or in terms of neighborhood school attendance centers; or, in terms of homeowners or tenants. All in a sense are clients with students being the primary client group. Further, there may be ad hoc kinds of sub-systems such as a group organized to attempt to bring about school reorganization, or to bring about a change in the school curricula such as a greater emphasis on vocational education.

It is also important to note that any individual can simultaneously be a member of more than one sub-system of the school district social system. Figure 2 illustrates some potential sub-systems of the school district showing some overlap of sub-systems.

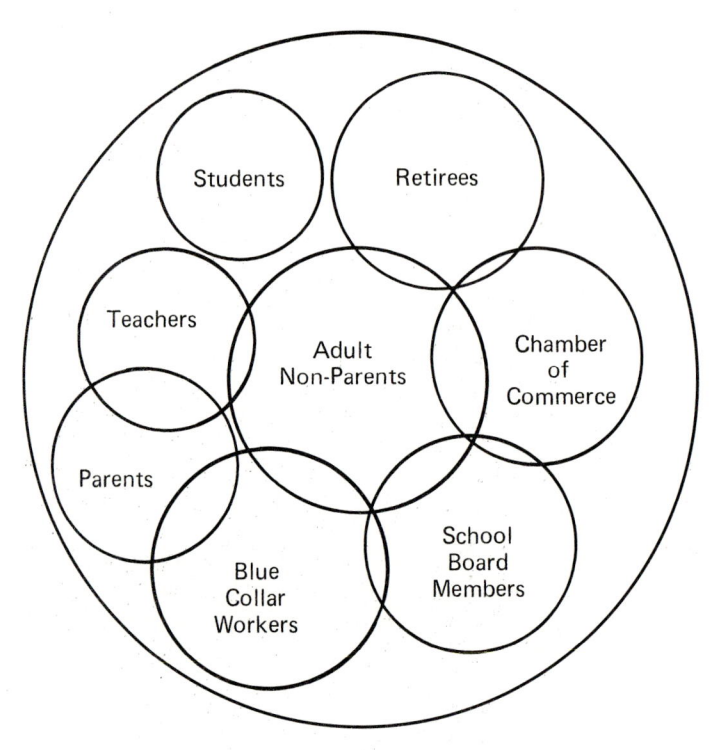

Figure 2. School District and Some Potential Sub-systems

Exchange of information and interaction occurs within and between the many sub-systems of the school district particularly as the information relates to the school district. Interaction is multiple and complex. The frequency and intensity of interaction within and between the sub-systems, and in this case within the supra-system school district varies a great deal. It must also be recognized that there is an exchange of information to and from the sub-systems of the school district and its supra-systems.

In summary, the school district in this chapter is viewed as a social system having both supra- and sub-systems. The school district is further viewed as a large, complex, pluralistic kind of social system responding in varying degrees of intensity to the information from both supra- and sub-systems. It is seen as having a very complex task in accomplishing the expectations and demands of its supra- and sub-systems and at the same time maintaining sufficient stability to survive and accomplish its purposes. The school district as a social system performing a designated social function must be dynamic and changing in response to its environment, yet it must be selective in its changes or suffer a complete loss of direction.

CLIENT SYSTEM

In the broadest sense, anyone who benefits from the services provided by a school district is a client. The most direct and tangible benefactor is the student. The least direct and most intangible benefactor is "society at large." This is not to say that "society at large" does not benefit. Clients are also contributors. Students contribute their time at the very least. In some instances they also contribute money as fees or tuition. Non-students, parents, and citizens contribute money in the form of taxes, and frequently they also contribute time as they serve as school board members, or on citizens' committees, or in some instances as volunteer teacher aides.

Each client can be viewed as having expectations for the schools. These expectations vary widely. Parents generally expect that their children will receive a basic education, that is, they will learn how to read, write, and understand simple arithmetic. They may also expect that their children will gain specific vocational competencies. They may further expect that values, concepts of morality and behavior, patriotism, citizenship, and social graces will be perpetuated by the schools. They expect the transmission of *their* culture, which usually means the transmission of elements of subcultures. Pluralism, therefore, compounds the problem of meeting the diverse expectations.

When the expectations of the overall society are examined, it becomes increasingly evident that in addition to the transmission of culture, schools are expected to play a major part in resolving social problems. Education and schools are seen by many as means to help in resolving social problems such as race relationships, poverty, violence, and unemployment.

All of the expectations from the many clients of the school system can be viewed as desired outputs of a social system. Students, parents, and those citizens with a less direct relationship to the schools expect certain outputs from the school. Their expectations are similar in some respects, and different in others. Nevertheless, the expectations are there, real and perceived, and realistic and unrealistic. In some instances, such as those that might result from a court decision, the expectations are imposed by a supra-system and not accepted or internalized by the local clients. *One task of the local school district in serving the client system is to determine the expectations of the clients as best they can. Secondly, the district must deliver the output to meet the expectations as best they can with the available resources.*

Just as clients expect output benefits from the schools they support, they also expect the benefits to be commensurate with dollar and time inputs. They expect cost-efficiency. They desire maximum output in return for their time and resources. They have a low tolerance for perceived discrepancies between their output expectations and their contributions. The problem of satisfying input-output relations is a complex one. The amount of output desired by various clients differs. For example, parents probably expect and demand more output than retirees. Yet, with the bulk of the educational monetary input in the United States still coming from property taxes it is possible that many clients who provide high input are not parents and perceive themselves as receiving very little direct and tangible output. This problem probably can never be completely resolved. Perhaps at best it can be compromised. Nevertheless, *a school district has the task of communicating the output-input relationship to the many client systems and reconciling it as nearly as possible.* In recent vernacular, schools are expected to be accountable; accountable for educational output, and inputs used to produce the output. To accomplish this they also need to be accountable for a client-school district information flow.

SYSTEMS ANALYSIS

The school district has been conceptualized as a social system with both supra- and sub-systems. Many client sub-systems, each with expectations for the school system, have been alluded to. Also the supra-systems have been recognized as having expectations. Information flow has been seen as a reality, whether it is accurate or inaccurate, and whether it occurs by accident or by design. Its frequency and intensity have been recognized as variable. It has been assumed that the school district as a social system is dynamic, that is as being responsive in some way or another to the information it receives and altering or changing its position in a selective fashion.

Clients have been viewed as being pluralistic in nature with varying expectations for the schools. They have also been conceptualized as contributing in different amounts to the enterprise, and as expecting little discrepancy between their output expectations and their contributions. It has been recognized that in all probability little likelihood for complete client satisfaction exists. Major tasks in serving the client systems have been identified as (1) determining client expectations, (2) delivering output to meet the client expectations as efficiently as possible, and (3) communicating input-output relationships to the client systems.

Systems analysis is a problem solving process. It attempts to analyze the overall operation of a complex organization as the organization seeks to accomplish its goals. The organization is seen in terms of its various components in interaction. The initial step in systems analysis is to clearly identify the problem with all its manifestations. Second, resources available to solve the problem must be identified. Third, a number of possible alternative solutions are generated and evaluated as to their likely success. Finally, a plan for problem resolution or reduction is selected and attempted, with feedback gathered to assess the relative success of the plan. Constant and regular monitoring is the rule with changes being made in the plan as it becomes apparent that they are needed.

A SYSTEMS APPROACH

In serving the client system, a primary task is to identify the output expectations of the clients. A data base is needed. There are at least two initial sources from which to gather output expectations; (1) the ongoing school organization, and (2) the clients.

It can be assumed that the existing school district is generating output. Upon inquiry, one could learn the extent to which a district had actually spelled out in writing their expected output. The extent of expressed, written output could range from nothing through a broad philosophical statement and rather broad goals to detailed performance objectives for all instructional and instructional support programs. If actual documentation is limited, then it becomes necessary to begin to identify goals at various levels. A way to do this is to identify and categorize activities that a school district is engaged in. Examples of instructional activities might include those dealing with elementary, secondary, and special education. Further, breakdowns of elementary activities might include subject content areas such as reading, arithmetic, science, social studies, music, physical education, and art. A similar breakdown could be made for secondary and special education. Instructional support activities might include administration, business services, guidance services, lunch services, and transportation. Decisions need to be made to identify and categorize activities. Further, decisions need to be made as to the level at which activities will be analyzed. For example, perhaps the broadest feasible categories might be elementary, secondary, and special education. On the other hand, the epitome would be to gather data on and for each child. A reasonable starting level seems to be the subject matter level by elementary and secondary schools. In some districts, heterogeneous in the nature of its clients, data may be desired by the attendance center. Once activities have been identified and a level of entry decided upon, it becomes necessary to generate specific performance goals or outputs for the activity. Performance goals must be measurable so that attainment can be reported. The generation of performance objectives is primarily the function of classroom teachers with the help of instructional support services. The performance objectives so generated represent what those employed by the organization perceive as being reasonable expected outputs. Further, the objectives so generated represent predetermined objectives by adults for their students.

While the organization is busy analyzing, producing, and categorizing its output, it is desirable to activate a citizens group to generate their perceptions of desired outputs. The citizens group can work best at developing broad overall system goals, rather than specific performance objectives. Also, the citizens group should be as representative of the population of the school district as is possible. In other words, a pluralistic population should have a pluralistic citizens group. It will be necessary to provide leadership to work with citizens to help them accomplish their task.

At or near the completion of the work of both groups, their products should be compared. Specific school district performance goals need to be matched to overall broad goals generated by citizens. It is possible at this time that there will be broad goals with no objectives, or objectives for which there are no broad goals. Discrepancies, if any, need to be reduced and reconciled.

The overall exercise, both within the organization and external to it, should result in rather specific, desired, measurable outputs for the school district. In other words, an expected output data base has been generated using professional employees of the organization and citizens.

A spinoff of the output generation activity can be a cost categorization or input system. As decisions are made on the categorization of activities and the level of activities at which records will be kept, in reality, decisions are being made for cost accounting. The cost accounting system should be in harmony with the output data system. In fact, a program budgeting procedure has begun to be identified. Refinements will be necessary, but by starting with expected inputs and doing some careful categorization, a data base is being generated for future programmed budgeting. Program budgeting generates data for decision making for the school district's administrators and for reporting to the client system.

Once output (performance objectives) and input (cost accounting) data have been generated two-thirds of a simplified systems model has been completed. The third component deals with methodology or process.

In instructional programs "process" refers to the instructional technique used, for example, self-contained classrooms, team teaching, individually prescribed instruction, or many other possible approaches.

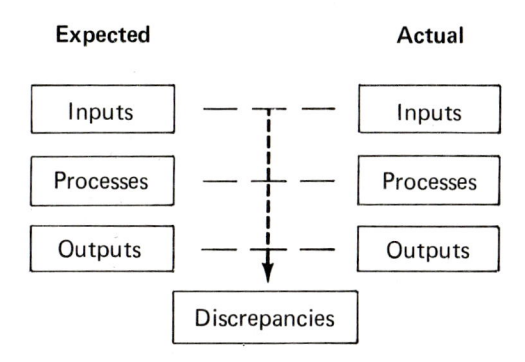

Figure 3. A Simplified Systems Model*

The first cycle of the system is portrayed in Figure 3. It is possible to report to the client system for example the inputs (costs) and outputs (achievements) of the many programs it operates. It is further possible for the school district to begin to analyze its effectiveness as it plans for the future. Once the actual data base has been established, input, process, and output can be manipulated or altered in an effort to achieve the most effective use of resources. They also have data that can be used to establish priority programs. For example, the community may, in times of austerity, demand a reduction in educational expenditures by refusing in one way or another to make monies available. The school district can respond by asking the community for their designation of low priority programs. The school district also, if they have maintained an adequate data base, should be able to predict the effects of an input cut on expected outputs for the various programs. It is *not* the position of the author that a planned-program-budget system as presented should be utilized as a defensive technique, rather that it be utilized as an analytic technique to assist administrators and clients in decision making. It perhaps is more effective as an offensive technique.

* This model is termed SPECS, School Planning, Evaluation and Communication System. It was developed by Terry L. Eidell and John M. Nagle of the Center for the Advanced Study of Educational Administration (CASEA) at the University of Oregon. During its development it was known as DEPS, Data-based, Educational Planning Systems. Further information on the model can be obtained from SPECS, General Learning Corporation, 5454 Wisconsin Avenue, Washington, D.C. 20015.

The model presented in Figure 3 represents a systems planning approach. Once actual data have been collected and the data base established planning can be further enhanced. Projected inputs, process, and outputs can again be designated. As time passes data should be gathered to compare how actual inputs, processes, and outputs compare with the expected or planned inputs, processes, and outputs. In other words, an evaluative feedback mechanism is put into operation. Discrepancies can be detected, explained, and corrected, and another cycle run. The technique should become increasingly refined.

The systems model presented is not a complete remedy for resolving all problems with a client system. It does, however, begin to solve a most crucial problem—that of generating and maintaining a sound educational data base. This data base, providing hard data on output, process, and input, provides information not only for decision makers, but also information that can be disseminated to clients. It should assist in regaining or establishing confidence in the schools. Further, it should result in improved educational services to the major clients, the students, who by many experts are considered to be the most effective public relations agents that schools have.

INFORMATION SYSTEM

Three tasks of a school district in serving its client system have been identified: (1) determining the expectations of the various clients, (2) delivering output to meet the expectations as effectively and efficiently as possible, and (3) communicating the output-input relationship to the clients. As a starting point it was suggested that the school district take initiative in establishing a data base. Since the primary clients are students, and the benefits to them are direct and tangible, it was suggested that the output data base be generated in terms of instructional objectives, and that these objectives be stated in measurable terms. While broad information in respect to desired output was gathered from citizens, the specific output expectations were generated from teachers. The school district was also seen as being responsible for generating process and cost-input data. These efforts by the school district are seen as basic and necessary to serve the client system. They do, however, represent only a beginning—a data base upon which to build.

An information system is intricately involved in the three tasks a school district has in serving its client system. The information system can be thought of as a non-instructional support system of the school district. As such it can be conceptualized in terms of the general systems model presented in Figure 3. It can be expected to produce certain outputs, utilize certain processes and require inputs (costs). An information system has at least four activities: gathering, categorizing, storing, and disseminating data. Each of these activities has expected outputs, processes, and inputs—the total of which represents the output, process, and input of the information system. Each of these activities will be briefly described as a part of a systems approach later in this chapter. At this time, however, it is pertinent to consider the task of the information system in respect to its environment.

In order to conceptualize the task of the information system it is helpful to recall the nature of the school system as a social system. Let us assume that the school district has one sub-system which is identifiable as employees. This group would include administrators, teachers, clerks, custodians, cafeteria workers, bus drivers, and anyone else officially employed by the school district. They are identified as being primarily in the internal environment of the organization. The external environment consists of all other persons residing in the school district that are not employees. It should be noted that employees do not interact with members of the external environment. Three sub-systems of the external environment are particularly important; students, school board members, and parents. Students, while categorized as being in the external environment, spend many hours of the day in contact with the internal environment and they cross the artificial boundary daily and transmit a large amount of information. School board members are members of the external environment who have been specifically designated to be responsible for maintaining a school system. They also are likely to interact regularly with the internal environment. Parents are likely to interact more frequently with the internal environment than are other citizens. Other citizens, as a large sub-system, also interact with the internal environment but probably with less frequency and intensity than students, school board members, and parents. Developing and maintaining an information system is viewed as a responsibility of the employee internal environment. However, members of the external environment also interact with the internal environment frequently and with varying degrees of intensity, and serve as an integral part of an information system. Employees, as they interact with the external environment, also are an integral part of the information system. This very important inherent facet of information flow in a school district deserves major attention.

It should be noted again that an information system includes *gathering information from clients as well as disseminating information to them.* Too often school districts seem to concentrate their energies on a flow of information to the clients. They advertise in formal ways what they are doing, maximizing their successes, and minimizing their failures, and in so doing often reduce their legitimacy in the eyes of their constituents. A gap begins to develop between what school authorities say they are doing, and between what clients perceive the schools are doing. A two way flow, in which school districts not only listen, but actively and systematically solicit information from clients can do much to improve communications. The information gathered from the clients helps to determine what information is important to them. Thereby the information disseminated is more likely to be directed to the clients' need for information. Also, the process of actively soliciting information in and of itself is likely to build confidence which enhances legitimacy. Let us now consider each of the activities of an information system in terms of the general systems model.

One major activity of an information system is gathering data. The data gathered represents the output of the activity. These data could be considered in three categories: (1) demographic, (2) attitudinal, and (3) those data related to the instructional output-process-input systems. Demographic data includes information such as population trends, employment trends, occupational information, socio-economic backgrounds of constituents, ethnicity, and level of education. It includes the kind of information gathered by the Bureau of the Census. In school districts that are experiencing demographic changes because of population mobility, a current record should be maintained. Existing computer capabilities make storing and retrieving these kinds of data by local school districts feasible. These data can be helpful to decision makers in anticipating outputs expected by the clients. For example, a decrease in actual farm employment in a rural area should be a signal to educational decision makers to alter the nature of its agricultural program. It may indicate the need for vocational programs aimed at employment in agricultural related activities. By the same token, a rural community that is changing to a suburban community needs to assess its educational programs to more adequately serve its clients. In the urban environment as the clientele changes it could very well be appropriate to offer English for Spanish speaking children, or early childhood programs to accommodate both the children and their working mothers. The trends revealed from demographic data can enable a school district to plan and respond to client needs and expectations as they arise. Too often educational programs lag behind needs. There are of course many reasons for the lag, however, a solid data gathering procedure and data base should at least enable a system to anticipate needs before they are glaringly and blatantly pointed out to them by irate, dissatisfied clients. Demographic data also provides the information necessary for such customary planning as building needs.

Illustrations of expected output expressed in performance terms from the data gathering activty of the information system in respect to demographic data might include the following:

1. By January 10 of each year the information office shall provide a revised pupil population projection for the next school year by grade level for each attendance center in the school district with ninety-five per cent accuracy.
2. By January 10 of each year the information office shall provide five year pupil population projections by grade level for each attendance center in the school district.
3. By April 1 of odd numbered years the information office shall provide a report giving percentage data on the occupations of parents of students by attendance center, in four classifications: professional, business and clerical, skilled labor, and unskilled labor.
4. By April 1 of each year the information office shall provide a report giving percentage data on the ethnicity of the total population of each attendance center in five classifications: white, black, Spanish-American, Oriental, and others.

The list is not exhaustive, it is merely illustrative of ways to express output expectations of a demographic nature.

Attitudes of clients toward education and the school district also provide a reservoir of valuable information to the school district. Attitudes form the basis for clients' expectations for education. Some clients, for example, might cherish or place a high value on foreign languages in the secondary school because of their culture and past experiences. Others may feel that work experience or non-paid charitable service is of great value. Still others may place little value on cognitive or psychomotor skills, stressing the affective domain of feelings and attitudes as being of primary importance. Illustrations of expected output expressed in performance terms from the data gathering activity of the information system in respect to attitudinal data might include the following:

1. By April 1 of even numbered years the information office shall provide a report giving rank ordered preferences of a random sampling of adults of the school district in respect to the instructional programs offered by the secondary schools classified by attendance center.
2. Within a period of three months after the conclusion of an experimental instructional program the information office shall provide a report

giving scaled responses of a random selection of parents of children in the program in respect to the success of the program.

3. By July 1 of each year the information office shall provide a report based upon structured interviews with selected persons in the community and selected persons within the employee group of the school system as to the perceived successes and problems of the school district.

Again, the list is by no means exhaustive, it merely illustrates ways to express output expectations dealing with gathering information of an attitudinal nature.

Output data from the information gathering component of the information system in respect to instructional system has been previously discussed. The information gathering component in terms of the general systems model must also consider processes. Processes refer to the specific methodologies planned to accomplish the expected output. Demographic data, for example, can be gathered from census materials and through self-designed survey forms for the local area. More difficult data to gather with objectivity is the attitudinal data. Techniques that could be used include mail surveys utilizing random stratified samples, interviewing, and open meetings. It should be remembered that the purpose of utilizing these procedures is to gather data, however, secondary side benefits are likely to be the opening up of communications channels, particularly those going from clients to the school district, and the attitudes that clients may develop feeling that the bureaucracy is listening to them.

Survey questionnaires need to be carefully designed so that they are clear, will obtain needed information, and can be quickly completed. It is the opinion of this writer that most of the time the survey questionnaires should address themselves to broad issues or goals. For example, the school district officials might want to know the feelings of various subsystems of the community on the relative importance of various instructional programs. These kinds of data can aid in decison making, and at the same time leaving school officials with the prerogative of making specific decisions. Each district would need to gather data pertinent to their unique environment. Random sampling, while not perfect because of incomplete returns, will, however, yield better data than hit and miss conversation. If it is stratified and categorized it can yield information by sub-system. For example, parents in the Sandburg High School attendance center might favor increased opportunities for vocational education, while those in the Harris High School attendance center might express satisfaction with the existing programs. Black parents may express a strong desire for early childhood centers, while white parents may be only mildly interested. Citizens throughout the school district, parents and non-parents may moderately favor early childhood education. While the collection of such data may reveal problems and cause some consternation, the lack of specific knowledge of the problems does not mean that the problem does not exist. Regular collection of survey data, by random sampling of sub-systems is a necessary part of a viable information gathering system. It is one of the few ways that information can be obtained from any constituents who do not through their own volition contact the school authorities. It permits anonymous responses.

Other means of gathering attitudinal data include interviewing, conducting open meetings, utilizing data gathered by employees in their regular patterns of interaction with clients, and data provided by various local advisory councils. Interviewing can be selective, obtaining much the same information gathered from surveys, but obtaining from specific selected persons. For example, valuable interviews may be conducted with business leaders, clergy, and labor leaders or other selected sub-system leaders. The interview being somewhat more open-ended than the survey, and also permitting two way communication, may be fruitful in obtaining additional information. Open meetings, conducted by both board members and administrators provide a forum for data gathering. It is good practice to schedule such meetings in various attendance center areas of the school district. Data gathered from such meetings should be evaluated for what it is, the opinions of persons probably representing an identifiable sub-system. These sub-system groups will likely favor their own values, or their own desires. Data gathered by employes can also be utilized. For example, as teachers conduct parent-teacher conferences, pertinent information can be recorded. Advisory groups also provide a vehicle for gathering information. A large school district may have reason to have a number of advisory groups. They may have one group responsible to the entire school district such as that discussed earlier in terms of generating overall goals for the school system. They may also have other specific advisory groups, such as student advisory groups, and those who deal with specific programs such as vocational education.

In summary, there are many sources in a school social system from which to gather data. Various processes of gathering data can and should be used. The data gathered should be as objective as possible and full cognizance must be given to its source. Random samplings from the entire community, for example, provide data representing the overall community, while an interview conducted with a labor leader, or the data gathered from an open meeting is likely to represent a sub-system. Both kinds of data are important.

The information gathering components also needs estimates for its cost of operation (input). These estimates can be based on time and materials which are both related to expected outputs and necessary processes to accomplish the outputs. With expected outputs, processes, and inputs identified, realistic decisions can be reached regarding the extent to which the information gathering system can be implemented.

Categorizing activities also can be analyzed in terms of the model. The following output goals are illustrative of the kind that could be applicable to the categorizing component.

1. Within a period of one month after receiving objective data, the data will be coded to meet the previously agreed upon specification requirements of the data gathering and data disseminating components.
2. Within a period of three months after receiving subjective data, the data shall be quantified and coded to meet the previously agreed upon specification requirements of the data gathering and data disseminating components.

Processes for coding will need to be designed. A coding system could be developed which would be patterned after the general systems model. Data could be classified as relating to one of the three major headings and then identified by sub-system source. This kind of classification would permit data to be stored and retrieved, for example, in terms of parents' expectations for educational output, general citizenry for input, or students' desires for teaching processes. This classification would be most appropriate for attitudinal data. Demographic data coding could be treated in a more traditional fashion. Input costs would also need to be calculated for the data categorization component.

The storage component is closely related to the categorization component. Output goals for the storage component depend heavily upon the equipment available. The goals generally will relate to rate of retrieval. For example, with sophisticated computer hardware, it is possible that data could be retrieved in minutes. With other kinds of equipment retrieval time could be estimated in terms of days or weeks. Also, the kind of data desired may complicate retrieval, particularly if further coding becomes necessary. As with the data gathering and categorization components, process and cost input also need to be projected.

Dissemination activities relate closely to and are dependent upon the data gathering, categorization, and storage systems. Dissemination of data involves providing pertinent and relevant information to sub-systems through an effective media agent. Two sub-systems have been defined; the internal environment, which is made up of employees, and the external environment, which is made up of all other persons in the school district, all of which in one way or another are clients. In the external environment a number of sub-systems can also be identified such as parents, retirees, and various occupational groups. Further, unique client groups such as students, school board members, and parents have been cited because of the amount of time they spend in the internal environment, the frequency with which they cross the artificial boundary, and the intensity with which they identify with the internal environment.

Information must be disseminated to both the internal and external environments. Employees need to know at least as much about the school system as clients. They need information for their own private benefit, but they also need it so they can transmit accurate information to clients as they interact with them. Employees are primary face-to-face type, powerful media agents. A plan must be developed to transmit information to employees. The plan must include more than the usual faculty and employee orientation procedures. Administrators at all levels need to identify their own functional roles in information dissemination. As a group they need to determine their coordinated function in dissemination activities in the internal environment.

Dissemination of information to the external environment has many of the same outputs and processes that are appropriate to the internal environment. The task is a bit more complex because of the many sub-systems in the external environment. In either instance, the overall goal is the same—to provide pertinent and relevant information to the audience. With precise data available, it seems most appropriate to be able to provide specific information appropriate to the audience in meaningful ways. For example, in the internal environment in respect to learning achievements it seems reasonable to assume that reports to the board of education would not need to be as detailed as reports to a building principal or a department head. By the same token, reports to the external environment should probably not be in extreme detail or reported in technical or pedagogical language. Illustrations of expected output expressed in performance terms from the dissemination component of the information system might include the following:

1. Analyze technical reports received from the data gathering component and prepare statements to be disseminated to specific subgroups in both the internal and external environment.
2. Prepare a newsletter to be distributed bi-monthly to all employees of the system.
3. Submit news releases to the media on topics of current interest.
4. Maintain a file of statements released, newsletters distributed, and clippings from local newspapers.

5. Arrange interviews for school personnel with appropriate news media.
6. Maintain a "hot line" or "rumor station" to receive messages and accurately answer inquiries.
7. Maintain a record and report regularly to designated authorities on information received from "hot line."
8. Prepare a monthly newsletter to be distributed to all patrons of the school district.

The list is not exhaustive. It is merely illustrative of ways to express output expectations of the dissemination component of the information system. Dissemination outputs by their very nature are process oriented. Process also includes the methodology to be used in preparing statements, newsletters, and other reports. Personnel need to have specific assignments. As in other components costs need to be identified.

The outline presented in this chapter for an information system while not detailed does represent a comprehensive systematic approach. Its implementation would require a staff of fulltime personnel. It is not likely that it could be successful if entered into as a "hip pocket" operation.

CONCLUSIONS

Serving the client system was conceptualized as (1) determining the educational expectations of clients, (2) delivering output to meet the client expectations as efficiently as possible, and (3) communicating the input-output relationships to the client system. Clients were seen in general as those who benefit and contribute to the school district. They were viewed as a part of a pluralistic social system responding to supra-systems and sub-systems. Major clients were considered to be students. An output-input systems model was proposed along with implementation suggestions to better satisfy the expectations of the student client subgroup.

Serving clients also requires an information system to learn their expectations, the characteristics of the school district, and provide information to the many kinds of clients. The output-process-input model was presented as a tool to plan and implement an information system. The information system was presented as having at least four interrelated components; gathering, coding, storing, and disseminating data. These components were analyzed and illustrative outputs and processes were suggested.

The topic was approached from an ideational overview framework, rather than from a detailed, precise, operational orientation. The framework, however, does have operational and implementation possibilities.

SELECTED REFERENCES

Alioto, Robert F., and Jungherr, J. A. *Operational PPBS for Education.* New York: Harper and Row, Publishers, 1971.

Andrew, Gary M., and Moir, Ronald E. *Information-Decision Systems in Education.* Itasca, Illinois: F. E. Peacock Publishers, Inc., 1970.

Clabaugh, Ralph E. *School Superintendents Guide: Principles and Practices for Effective Administration.* West Nyack, New York: Parker Publishing Company, Inc., 1966.

Eidell, Terry L., and Nagle, John M. "Program Planning Document for Data-Based Educational Planning Systems." Paper developed at the Center for the Advanced Study of School Administration, University of Oregon, 1970.

Hartley, Harry J. *Educational Planning-Programming-Budgeting.* Englewood Cliffs, New Jersey: Prentice-Hall, Inc., 1968.

Lyden, Fremont J., and Miller Ernest G. *Planning Programming Budgeting: A Systems Approach to Management.* Chicago: Markham Publishing Company, 1968.

McCloskey, Gordon. *Education and Public Understanding.* New York: Harper and Row, Publishers, 1967.

McGrath, J. H. *Planning Systems for School Executives.* San Francisco: Intext Educational Publishers, 1972.

Morphet, Edgar L.; Johns, Roe L.; and Reller, Theodore L. *Educational Organization and Administration.* Englewood Cliffs, New Jersey: Prentice-Hall, Inc., 1967.

Owens, Robert G. *Organizational Behavior in Schools.* Englewood Cliffs, New Jersey: Prentice-Hall, Inc., 1970.

Pfeiffer, John. *New Look at Education, Systems Analysis in Our Schools and Colleges.* New York: Odyssey Press, 1968.

Thomas, J. Alan. *The Productive School: A Systems Analysis Approach to Educational Administration.* New York: John Wiley and Sons, Inc., 1971.

10

JOHN W. GILLILAND
DARWIN WOMACK

Trends In School Plant Design: Facilities for Today and Tomorrow

Of all the activities in which American people engage, while living and working together, perhaps none expresses in material form so many aspects of our culture as schoolhouse construction. The growing national concern for the improvement of education is undoubtedly one of the greatest phenomena of our time. Dimensions of education: purposes and content, methods and materials, as well as financial support and status, have beeen dramatically influenced during the past decade by forces inherent in rapid cultural change. Educators in every corner of the nation have examined the educational process with the full realization that the future of this nation and the free world depends to a considerable extent on excellence of education.

Yet it is depressing to note that much of the money spent for school construction in recent times has gone for facilities out-dated years ago. Many schools have the same basic floor plan as did the school buildings of the 18th and 19th centuries. Architects and educational leaders have come to realize that a school facility must be designed to house and implement the educational program. Since the program is changing, the building of the future must be designed in a way that the physical elements or parts of the facility do not hinder the performance of pupils and teachers. All blocks and barriers to optimum performance must be removed. Gone are the days when a new structure is built without involving pupils, teachers, and laymen. Today's teachers, administrators, boards of education, architects, and laymen must work together in developing a functional school building.

EVOLUTION OF SCHOOL BUILDINGS

The entire history of the development of school facilities reveals an effort to provide an indoor climate to which man is best adapted. School buildings have generally changed slowly but are always seeking the ideal indoor climate. The founding fathers realized that an educated public was necessary if a democratic government was to function in the proper manner. Buildings to house the students had to be provided, and for some time, attention was given only to the kind of roof and walls in a school building. Furnishings consisted of a rough hand-made bench and stool. During the early colonial period, the schoolmaster's home was used, often the only building available for teaching boys and girls. Light and air came through a small opening in the wall. The one-room school building was the first facility used for educational purposes. This rectangular-shaped structure paved the way for the school buildings that we have today. The one-room school featured one teacher for all grades, a pot-bellied stove, a few windows, oil lamps for light at first, but later one electric light, hanging from the center of the building, providing the only light for the classroom.

The Lancastrian system of education was introduced in the early 19th century. A teacher taught fifty monitors who each taught ten students, using one large room for instruction. The room was usually about 50 by 100 feet for students or ten square feet per pupil. Benches and tables were the only furniture and equipment available. Comfort of pupils and teachers was not considered as being important for the learning process. The Lancastrian system had something to do with establishing the idea of public education for all pupils, thus helping to convince the public that a system of public education should be extended and given financial support at the local level.

One-room schools were prominent for many years and a few are still being used today, however, the trend has been in the direction of larger school units, housing more and more pupils. Large enrollments made it necessary to have a plan for grouping, hence the graded type of organization found in most schools today, came into existence. Change in the school design was necessary to accommodate the new type of organization that was used. The Quincy School of Massachusetts is an example of architectural change in a school building. The Quincy building had a basement, a top floor assembly hall, and twelve classrooms with an area of approximately 800 square feet in each room. Enrollment averaged about fifty-five pupils for each room.

Several changes were made in educational programs in the latter part of the 19th century. The Kalamazoo Case in 1872 and other cases settled the issue of public support for schools. Manual training patterned after European educational programs was introduced in 1878. Building changes corresponded to new curricular offerings in such courses as machine shop and woodworking. Laboratory methods of instruction came gradually into being during this period. John Dewey's influence, affecting the approach to teaching, developed in the early part of this century. The newer educational programs called for greater involvement of the learner and larger spaces for teaching. More creative teaching and using the out-of-doors as an environment for learning stimulated changes in buildings, giving greater emphasis to planning a facility to implement the educational program. Provision for greater flexibility of teaching spaces, better lighting, improved furniture and equipment are all emphasized. More and better physical education programs were demanded at the onset of World War I. New approaches to education stimulated change in facilities, emphasizing that a building is designed to implement the educational program. Opening up the facility on the inside, providing for adaptability of spaces and for large and small group instruction affected school building design.

The depression of the 1930's slowed school building, but postwar growth caused an increase in the number of buildings. Standards for buildings were developed following the war and instructional programs were improved. New courses required additional space so the educational program could be carried out in a manner that was best for the students. Provision for flexible space was limited to non-load bearing movable walls with a small amount of movable furniture. Following World War II, building programs gave rise to even greater emphasis on flexible spaces. Folding partitions or movable walls were used in room dividers in the 1950's. Visual dividers were introduced in the 1960's.

Changes in educational programs, philosophy and methodology during the 1960's demanded greater flexibility through open space classrooms utilizing furniture as visual dividers in teaching spaces instead of sound retarding, operable walls. Doors and permanent walls were eliminated. Visual divider walls made it possible to arrange furniture and equipment in a way that was suitable for most any size group. Considerable attention has been given to physical conditions in the classroom that are necessary to provide optimum conditions for learning. The main idea was to do away with blocks and barriers to learning inside the facility. Many changes in design occurred due to demands related to changes in the educational program.

CURRENT INNOVATIONS AND TRENDS

Change in school buildings in the 20th century has moved from a series of one-room schools to a facility housing hundreds of pupils. Classrooms are air-conditioned with a more precisely controlled environment and flexibility. The circular polygonal buildings have helped to break up yesterday's sprawling finger type and at the same time provide a more cheerful and carefully controlled environment enhancing achievement of pupils.

Compact buildings reduce the perimeter without decreasing area inside the building. The demand for more machines for the use of pupils and teachers, provision for greater flexibility, and sound control are easier to obtain in a modern school due to the fact that these items may be included without the building costing any more than a conventional facility. The kind of teaching spaces determine, to a considerable extent, how well the educational programs may be implemented. It is when you move inside that you make or break a school building.

One cannot help feeling the change in tempo of our educational programming. Perhaps the most distinguishing characteristic of anything living today is that it is forever changing. A school facility that never changes gives a good indication that the program within is dead, or was never alive. Therefore, in order to provide for a living, dynamic, and effective educational program, a school plant containing the kinds of teaching spaces needed to implement the program must be provided.

The ferment of ideas in education directly related to planning a school plant, provision for flexibility, scientifically designed furniture, proper equipment, good thermal and acoustical control, and good classroom lighting are essential in a school building, if a good environment for learning is to be achieved. Since school buildings have such a direct and definite influence on the achievement and progress of boys and girls, educational leaders are concerned about designing a building to implement the educational program. The building must be planned to serve pupils, teachers, and community providing an optimum environment for learning.

School building design is changing for reasons of economy. With spiraling construction costs, there is a trend toward the compact building which reduces the perimeter. Brick-masonry, perimeter walls, one story in height, cost approximately $150 to $175 per linear foot. Up to $250,000 has been saved on one school building by reducing the perimeter, reducing size of roof area, shortening of piping and ducts, as well as reducing the size of corridors. Large amounts of glass and interior spaces are also being reduced more than 25 per cent. Reducing the cost without decreasing the quality of the building saves dollars to spend on the inside of the structure.

Special concern is focused on the role of flexibility in school planning, especially as an important element of the learner's physical environment. The open plan school has the capability to implement modern educational programs as well as programs in the future. Providing for flexibility in the classrooms makes the spaces more adaptable and functional. Flexibility exemplifies mutability and hits at the heart of one of the barriers that gets in the way of providing good physical conditions for learning. Provision for flexibility in a school makes it possible to adjust to new situations or conditions that may arise in the future. The need for flexible school facilities has been demonstrated conclusively in numerous instances. To obtain the greatest return on the educational investment, individuals concerned with the educational well-being of pupils need to be cognizant of the potential of opening up classrooms inside the facility.

The open plan school offers a way to implement the educational program and create favorable conditions for learning. Open spaces dictate the use of movable walls or visual dividers which may be in the form of movable bulletin boards, tack boards, chalk boards or storage cabinets. The use of dividers may increase utilization of the area in order to provide a more desirable teaching and learning environment. The use of school furniture for other purposes such as visual dividers, cuts costs. Open wedge-shaped classrooms, high utilization of spaces with operable interior walls and adaptable furniture are no longer a dream but a reality. There is considerable evidence of the many fine things that may be accomplished through large and small group instruction accomplished by opportunities for individualized study. Research indicates that some subject matter may be taught as well or better in groups of 150 than in groups of 25. Successful instruction depends upon the teaching techniques used. In one school system, typewriting for personal use, taught by one teacher to groups of 120 provided better results than five teachers with 24 students each. The very nature of changing society demands changes in school buildings. Spaces for teaching need to be designed for flexibility, serviceability and adaptability, if the present day programs of education are to be implemented to the fullest extent.

There are problems related to the use of flexible or open spaces. These problems are related to the use of spaces by staff members who have spent most of their life teaching in self-contained classrooms. These teachers have been taught largely by the lecture method. Special assistance through in-service education programs are provided in many schools that may help the teacher adapt to modern methods of teaching. There are many ideas that may be called to the attention of the open space teacher. Absorbing sound waves through the use of carpet as a floor covering, speaking to pupils in a way that the sound level is best, securing cooperation of children in keeping the carpet clean, using earphones for high sound level presentations and others help the teacher and pupils to keep the sound waves at a lower level.

School planners and educational leaders have discovered that the quality of instruction may be better in the open plan classroom. Functional furniture and equipment are used as visual dividers as well as for storage, chalkboards and bulletin boards, which result in lower cost for the building. Other savings accrue on the building by using a compact design. Carpet is installed in the classrooms providing for sound control as well as other important benefits that may accrue. All of these items are available as money is saved on permanent walls. The money saved on walls is spent for better furniture and equipment which help to provide better conditions for learning.

MEDIA AND MACHINES

Considering the vast amount of information available, the classroom teacher is confronted with greater responsibilities than ever before. If provision for machines is ignored, the building is out-of-date the day you move into the facility.

The utilization of educational media in schools across the nation is coming to mean far more than such familiar teaching aids as audio-visual equipment, programmed texts, and computers. In a broad sense, the emphasis is on technology in education; the engineering of the entire environment so that students may have favorable conditions for learning. Regardless of the methods used to bring about the best environment for learning, there is a deep commitment to discover those teaching and learning processes that are most effective, and how these processes may be aided by technology and the influence of diverse educational media. Perhaps the most promising aspect of utilizing various media is the possibility of new technology to assist teachers in doing a better job of teaching the individual student. Instead of lock-step methods of education with the teacher working with one classroom sized group, perhaps through the use of devices and techniques that facilitate self-instruction, the teacher will turn into a manager of learning resources for each student.

There are a number of reasons for the emphasis upon the use of media in instruction: (1) the population explosion has resulted in many additional persons to be educated; (2) the information explosion has resulted in an ever expanding amount of knowledge available; (3) new curricula with emphasis on methods of teaching through participation and experimentation; (4) an increased demand for quality education for all individuals; and (5) the demand for better utilization of teachers. These situations have placed a greater burden on educational media to assist in the instructional process and at the same time the teacher has had to assume greater responsibility.

There are many reasons why media utilization has not received the emphasis necessary to meet stated needs. Education has trailed far behind other sectors of society in utilizing modern technology. Some educators have felt that new media should be complementary and not supplant other sources of learning. Attempts are made to fit various media into a pattern that will support optimum utilization of the complementary function. Some feel that new media puts more distance between the teacher and pupil. If that were the case, the reason would be that media considerations were determining the curriculum instead of overall educational considerations. The long delayed deployment of media in education has finally given away to rapid and accelerated use of this type of instruction. This utilization necessitates numerous changes in educational facilities and curriculum such as new designs in school buildings, flexible class scheduling and other instructional procedures. The increase in the use of media creates the need for controlling airborne or impact noise, which may influence the design of the facility. Spaces of various sizes, special screens, and other electronic effects to accommodate audio and visual media are all needed to accomplish good conditions for learning. Real progress has been made in the use of media in modern buildings. Many schools have learning centers, teaching machines, closed circuit televison, study carrels equipped with electronic aids and other types of machines. Coaxial cable and other kinds of communication devices are installed in many of the newer school buildings although some schools may have installed the machines but never used them to any great extent. It has been found that it is difficult to find the technicians needed to keep the machines going. A technical expert is needed to operate the equipment, thus helping teachers to use the machines to the greatest advantage.

ENVIRONMENT FOR LEARNING

Educators and school planners have been concerned with thermal environment or temperature conditions in classrooms for many years. Teachers have attempted to control heating and ventilating by simply raising a window or turning up the heat. For a long time these simple methods of conditioning the air were the only ones available. Much insight has been gained through scientific studies regarding thermal environment and the effect of temperature on learning. Such studies have revealed that maintaining proper temperature environment in the classroom increases the learning and work efficiency of pupils and teachers. Proper control of the temperature environment in the classroom will affect the ability of a pupil to grasp instruction. Both working and learning efficiency decreases with departure from an optimum temperature which is 73 to 77 degrees for most school activities. If environmental conditions are not favorable, the pupils behavior will be affected. Pupils will not react in a practical manner but in a way related to physiological and psychological factors. In contrast to adult behavior, children will boil over with restlessness in a hot room, daydream in a cold room, seek the light in a dark room, whisper in an ultra-quiet room and compete with noise in a room with a high sound level. The effect of temperature on human activty is perhaps more important than any other factor. The basic reason for building a school is to provide favorable conditions for learning. There is a comfort zone at which the human body performs at the highest level. Authorities agree that simultaneous control of the thermal elements, mean radiant temperature, relative humidity, air movement, and air cleanliness is of primary importance in providing a favorable environment for learning. Authorities agree that as stress caused by improper temperature is reduced, fewer mistakes occur. Control of these elements is a requirement for adequate functioning of the human body. Most people believe that the human organism is highly adaptive yet it cannot compete with all of the blocks and barriers that seem to come from improper thermal control. Stress on the human body increases with more adverse temperature conditions. The highly complex body does not function well at high or low temperatures. The efficiency of the student is impaired when the condition of the air and surrounding surfaces affecting physical and mental comfort depart as much as two degrees above or below a temperature of 75 degrees.

Evaluation of temperature conditions inside a building is of the utmost importance, especially as temperature control is related to the physiological capital of an individual. Every person has a certain amount of energy that he uses in going through the various activities each day. Seeing is said to use 25 per cent of a person's energy. All of the activities pursued use all of the energy available, sometimes by eleven or twelve o'clock in the morning. Energy is often referred to as physiological capital. If we work in spaces that help preserve our energy, we not only last longer, but achieve at a higher level. A warm, sometims very warm, classroom affects the well-being of pupils and teachers, using up physiological capital much faster.

The effect of temperature on human activity is perhaps the most important of any single factor, although many other factors are important. There is a general comfort zone at which the human body functions with less stress and fewer mistakes. The individual may differ radically in his ability to function in areas outside the comfort zone. Some of the more exerting physical activities, such as that of a marching band, physical education activities and classes in typewriting will require temperatures in the lower range of the optimum zone. When the temperature changes more than two degrees from the optimum range, physiological stress is experienced. The amount varies as related to more adverse temperature conditions.

Due to differences in dress, in metabolic rate and in the reactions of individuals to heat and cold, opinions vary among occupants of a classroom concerning a desirable temperature. Men generally prefer a temperature about two degrees below that preferred by women. Pupils prefer a cooler temperature than do adults. In one study, 140,000 temperature readings were taken. It was found that forty-four per cent of all classroom temperatures included in the sampling were above 75 degrees, which is above the upper limit of an optimum temperature. Eighty-six per cent of the readings were above 72 degrees and 2.1 per cent were below 70 degrees, which indicates that classrooms are too hot most of the time. High classroom temperatures may have been due, in part, to the teachers who controlled the thermostat in a room. An overheated classroom accompanied by inadequate ventilation, leads to stress producing situations that cause pupils to be inattentive and restless which may in turn cause discipline problems. When the environment is too warm there is ample evidence to prove that the ability to perform various tasks deteriorates. An overheated child is prone to cease concentration on academic matters and relax into daydreams. An overly warm classroom may cause pupils to be less attentive, and make more errors in their regular classroom work. There is some indication that students may experience approximately 2 per cent reduction in learning ability for every degree that the room rises above an optimum temperature.

Relative humidity becomes a critical factor in determining the temperature to which a subject may be exposed and still maintain proper heat regulation of the human body. There is evidence to indicate that the effect of humidity on the human body and the thermally related atmosphere, has at times been underestimated. Relative humidity is related to and influences comfort. High humidity helps to create greater temperature stress. When the temperature in a classroom reaches a favorable level, the relative humidity is about fifty per cent, plus or minus ten per cent. Exterior walls and windows exposed to the elements make temperature and humidity more difficult to control. As a result of solar heat gain of 200 British Thermal Units per square foot of glass in a room, an excessive amount of window area makes adequate temperature control very expensive. An over-heated pupil is prone to cease or lower concentration on academic courses or daydream much of the time. Classrooms are extra warm most of the time. Heat is built up in the room through solar heat from the sun, shining through glass windows in the exterior walls. Heat is also generated from the lights in a classroom as 3.4 B.T.U.'s, are developed for each watt of electricity used. Pupils influence heat gain as each one brings 350 to 700 B.T.U.'s of heat into the classroom. A typical heat gain or heat load may total 70,000 B.T.U.'s in a single classroom. This is more than enough heat needed for a classroom. A system should be available to control humidity, air movement, and temperature at all times. This, of course, means air conditioning.

Educational leaders are giving intelligent thought to the selection of heating and ventilating equipment for schools. Some continue to plan and build buildings that impose a terrific burden or stress on students and teachers. School buildings should be designed for an optimum temperature of 75 degrees Fahrenheit, plus or minus two degrees, incorporating such considerations as minimum window area, covering on the window, and building orientation. If a school building is designed in a way that it may become a block or barrier to the pupils and teachers, the most important purpose of the building is defeated, as a facility is developed in such a way as to improve the environment for learning. The evaluation of the thermal environment, in terms of the most advanced educational thinking, with a temperature that is desirable for a high quality educational program is necessary to provide a good environment for learning.

The ear ranks second only to the eye as a corridor to the mind; therefore, the sonic environment in a school assumes great importance as related to the learning process. Unwanted sound or noise interferes with communication inducing discipline problems. Noise must be held to a minimum in order that pupils and teachers may hear. Optimum sound control must be provided if there is to be a good environment for learning. Many school buildings even today are ignoring certain types of acoustical controls and there is yet a widespread use of hard surface such as glass, glazed tile, asphalt and vinyl tile, steel, and smooth plastered walls that create waves of reverberation that interfere with acceptable hearing. A high sound level interferes with hearing, which affects the communication process.

The need for better acoustical planning is emphasized by many recent changes in education. Activity of many types is the cornerstone of many schools and the future appears to be filled with more machines and open space, for individualized programming and learning. Teachers are concerned about too much noise in the classroom. Controlling sound in the classroom is necessary if there is to be a good environment for learning.

The effects of improper sound control in schools are pronounced in two important areas. The first is that verbal communication is highly sensitive to the masking effects of many noises. In most classrooms speech is accompanied by unwanted sounds. It is a common experience to have syllables, words, or even sentences drowned out by noise. How much learning in the classroom is lost due to interference of noise is not known, but clearly in this case there exists a need for optimum speech intelligibility which is exceeded in no other situation.

The second major effect of unwanted sound in schools concerns the physiological and psychological well being of the individual. Reactions to noise have been observed to include (1) annoyance and other emotional reactions, (2) loss of efficiency in performance, (3) auditory and physical fatigue, (4) hearing loss, and (5) other reactions such as nausea, headache, and muscular stiffness. It is clear that such effects drain valuable physiological capital from individuals at times when such assets are most needed. Educators must be concerned about these barriers to learning and be willing to give support to proper design for controlling sound in schools.

Sound waves are relatively simple to understand. A sound source in a classroom sends spherical waves outwards but encounters many objects which gives rise to a very complex distribution of the waves. When a sound wave strikes one of the room boundaries, some of the energy is reflected from the surfaces (reverberation), some is absorbed by it (absorption) and some is transmitted through it (transmission). Objects in the room also cause reflection and absorption. The reflected sound is again partly reflected, partly absorbed, and partly transmitted. This cycle is continued until the energy of the wave is dissipated.

In controlling sound in spaces where speech communication is essential, the main task is to absorb most of the sound at the boundaries of the room in order to eliminate reverberations. Sound waves are best reflected from hard, flat and smooth surfaces. Many acoustical materials are now available that have high absorption qualities. The most effective sound retarder now in use in school buildings is carpet. This product not only absorbs sound waves but it almost eliminates noise which may originate from impacts with the floor. Carpeted areas are free from the usual sounds of walking or dropping of books, pencils, and other materials. Furniture and equipment may be moved from one position to another with less disturbance, thereby lowering the sound level. Carpet is so effective that spaces may be opened without causing undue stress or communication difficulties. In addition, this will save in cost of wall construction and improve the space as far as flexibility is concerned. Other sound absorbing materials are valuable when used as surfacing for walls and ceilings, as suspended units, as linings for barriers and enclosures for confining the noise of specific sources, and as linings to reduce noise transmission through ducts.

Air conditioning of the building is necessary for effective reduction or elimination of noise from vehicular traffic such as trucks, automobiles and airplanes. Isolation of shops, music, physical education classes and other sources of high intensity noise will also be necessary. Dense walls which provide for low transmission of sounds should be used when these conditions are in close proximity to areas requiring a reasonable low level of background noise. Mechanical equipment used in instructional areas should conform to low levels of noise emission. Since more modern schools are likely to have more equipment for instruction, these devices should be selected with care. In open plan schools, zoning or use of many audio devices will help eliminate unwanted sounds in large areas.

Large volume space, such as a gymnasium is even more difficult to plan because of problems with sound control. Band rooms and rooms with domes are also difficult to design acoustically. The educator must rely upon the professional designer but it should be pointed out that too often designers do not seriously consider sound problems unless the user points out the necessity for good acoustical control. Proper attention must be given to size and shape of a space and balance between hard and soft surfaces in order that undue stress on pupils and teachers may be eliminated.

The importance of proper visual environment cannot be overestimated. Over twenty-five per cent of a person's energy is used in the process of seeing. This represents a large expenditure of the body's physiological capital. Abuse of the human eye has many harmful consequences, such as the problem of eyestrain as related to glare. Illumination and brightness difference are important factors in planning classroom lighting. The visual environment is a three dimensional pattern of brightness and color visible to a person within the visual field. The visual environment also includes emotional and esthetic values which are less easily measured, but very important. Principles of lighting quality have been established which will provide a comfortable, efficient and pleasing visual environment. Proper lighting is necessary in order to avoid injury to the eye. The lighting environment influences posture and balance which may result in serious physical handicaps or physical deterioration.

An important concept regarding a good visual environment is that brightness differences in a classroom must be kept within recommended limits, such as a brightness difference ratio of three to one in the visual field and ten to one for the total area of the classroom. The angle at which light falls on an object is important, but surface brightness in the field of vision is of greater importance. Extremes in surface brightness are factors causing discomfort, loss of efficiency and physical deterioration.

Glare is a factor in the decline of overall accuracy, which affects an individual's perception. It is necessary to limit the brightness of sources of light exposed toward the work so that seeing is not hindered by reflection from the detail of the task or from the background. Proper school lighting produces a visual environment so that a person may see efficiently and effectively without distraction from any of the parts of the luminous elements of the seeing environment. Adequate levels of illumination, with properly balanced brightness, aid the educational process. Good classroom lighting reduces visual fatigue and helps create a cheerful and pleasant atmosphere. The goal of classroom lighting is to allow students and teachers to see comfortably and efficiently without undue glare distraction. An adequate, balanced level of illumination is necessary for the efficient performance of a student.

The use of color in the classroom is related to good classroom lighting. Color is utilized to gain certain ends and research indicates that color affects living things. Educational planners and architects, responsible for planning school buildings have often ignored color as related to its effect on learning. Color is a most important factor in good classroom lighting. The color of walls, ceilings and floors affect the quality of classroom lighting. The psychological and physical effects on human beings, including: overt emotional reactions, feelings of warmth and coolness, alteration of judgment, variable appetites, apparent camouflage, variable fatigue, accuracy in focusing the eyes, and illusions of change in size, weight or distance should be given adequate consideration in classroom lighting. The selection of colors for classrooms and other teaching areas should be individualized by recognizing the needs of older and younger pupils as well as where the classroom is located. The prime factor in color choice should be provision for an appropriate learning environment in order to enhance the mental, physical and emotional well-being of pupils and teachers. Students and teachers experience fewer visual difficulties in spaces that are decorated in a proper manner. The psychological and physiological welfare of pupils and teachers is improved through the correct use of color within the school. Maximum benefits from the use of color are obtained by careful study and application of the information gained. If colors are selected that are psychologically acceptable, and provide for proper reflectance students and teachers tend to utilize such spaces to the maximum.

Furniture and equipment for a new or rehabilitated building should be given consideration as this is a factor that is closely related to improving conditions inside the facility especially in the teaching spaces. Good planning in the selection of furniture and equipment provides a better environment for learning. Furniture and equipment should be selected on the basis of contribution to and compatibility with the total educational program. Provision for optimum learning by selecting the proper kind of furniture and equipment is a must in planning a good school building. Furniture and equipment are associated with the instructional function of a school. Furniture and equipment must be usable and functional by all school personnel. Furniture must be movable at will and with ease. Furniture should not only be flexible, but must be adaptable so as to meet the function of serving the needs of students, especially the individual differences of pupils. Administrators, teachers, and manufacturers are realizing more and more that comfortable, colorful, functional furniture and equipment contributes to a better learning environment. While it is true that shelter afforded by a building is necessary, the kind of learning that takes place inside must support the educational program. With the emergence of modern teaching methods and other program changes, new concepts must be utilized in selecting school furniture and equipment.

In the past, the emphasis in selecting and purchasing furniture and equipment was the low initial cost. The current trend is shifting to scientifically designed furniture of high quality as it costs less in the long run, simply because it lasts 25 years or more, while low initial cost furniture may not last more than four or five years. Criteria should be observed in the selection of furniture and equipment. The most functional and economical furniture in the long run is that which serves the school's purpose as intended, and consequently enhances learning. The furniture should: (1) be scientifically designed, (2) encourage good posture, (3) utilize reflective colors, (4) the reading surface of the furniture must have a proper reflection level, (5) be resistant to tampering and destructive efforts, (6) have rounded safe surfaces, (7) be flexible, and easy to move, (8) special features for noise reduction and/or control, (9) attractive, thus enhancing appearance of the classroom, (10) durable, and (11) provision for minimum maintenance. Several years ago when a school building was planned, all of the money available was spent on the new structure, thus moving old furniture and equipment into the new facility. This procedure minimized the importance of good furniture and equipment. What has been said regarding criteria for the selection of furniture is equally applicable to equipment. The library, classrooms, service areas and all spaces should be considered when purchasing furniture and equipment. A description of available space, a thorough study of the activities, plus a vision of the new and expanded programs are essential before proper consideration can be given to purchasing furniture and equipment.

The purchase of furniture and equipment is a long term investment and should receive careful consideration. School officials must remember that tomorrow's educational program will be greatly affected by the kind of furniture and equipment that is purchased today. When purchases are made for proper utilization, there is little doubt that the educational program may be carried out in a more effective manner. Pupils are the primary concern in purchasing furniture and equipment. It must have proper height, be comfortable as well as perform the function desired. More emphasis is being given to individualized instruction and research, and in order to be effective, pupils must have access to and be able to use many instructional aids. When equipment is a tool in the hands of the learner, it becomes a stimulator for creative effort and helps to fashion an environment where imagination may roam and learning becomes more exciting. Our changing philosophy of education now demands more flexible and movable types of furniture and equipment, which permits multi-use and convertibility of space.

If schools are to fulfill an important mission, consideration must be given to the best possible means of enhancing the learning environment. This calls for careful evaluation of furniture and equipment that is best for all situations. Careful planning and far-sighted vision on the part of all concerned can do much toward producing a healthful and productive learning environment. A dedication to this task is essential.

EDUCATIONAL SPECIFICATIONS

Most school administrators have begun to realize that before the design of a building is undertaken, a set of educational specifications should be developed. Educational specifications include statements regarding the kind of program to be carried out in a building and are desirable for the architect and educators connected with a school system. One of the first benefits achieved is more effective planning. In the past the architect for a school building was told to build a school with a certain number of classrooms. School officials and teachers did not become very much involved in planning a facility. Little involvement of the staff in planning results in a building that is not acceptable to the staff. When educational specifications are developed the primary focus is on involving teachers, pupils, and laymen. When the staff is involved in planning a facility, they are more receptive to modern ideas and innovations, thus the program potential of the building is realized more completely. As educators become more involved in planning, less wasted space is usually found in the facility. Who knows more about how much and what kind of space is needed than the teachers? When teachers and others have determined the kind of equipment needed, the architect finds it much easier to design a building that will implement the program.

Community leaders and lay people should participate, in so far as possible, in planning a school building. Those involved are more willing to finance a facility if the purposes of the school program are more clearly understood. No better way exists to let people know what is going on than to involve groups in the development of an educational program for a school. Others that may participate in the planning are educational consultants, architects, students, and occupational advisory groups. With all of these groups involved, the development of educational specifications is brought about through cooperative effort of many people.

Developing specifications related to the curriculum in a school may include what is generally spoken of as developing general and detailed requirements for the educational program. Under the heading of general requirements are found: (1) statement of philosophy and objectives of the educational program, (2) a plan of the proposed educational organization as well as the groups to be accommodated, (3) a description of spaces for the educational facility, and (4) consideration of furniture and equipment. Under detailed requirements, the school is broken down into many parts. Statements of philosophy, objectives, and trends for each subject or course offered are listed. The number and size of spaces needed are suggested as well as the manner in which the spaces may be divided. For instance, if visual dividers or movable walls are desired, this will be indicated. The listing of kinds of spaces are given for each area of the curriculum.

A very important advantage of involving the staff in the development of educational specifications for an elementary or high school is the involvement of teachers. Teachers and others may think that they know little or nothing about planning a facility in the begining, however, they soon begin to understand a good school building. Writing educational specifications are valuable in direct proportion to the degree of involvement and the ability to communicate ideas. The people who may teach in a space have an important contribution to make in planning. The teachers should be given the opportunity to make decisions and their suggestions should be given priority. Do they want open spaces, what kind of furniture and equipment do they want or size and number of teaching spaces should be considered. This takes time, perhaps four to six months of meetings. The development of specifications by involving teachers, requires help from the architect, the administrator and consulting specialists, all serving as resource people. It is important to remember that those who have a hand in planning a facility will come more nearly using the spaces inside the way spaces should be used.

SCHOOLS OF THE FUTURE

Until recently, the American schoolhouse seldom varied from the shape or form which had prevailed for centuries. Though there has never been a national system of education, facilities throughout the nation look so much alike as if there had been a legal requirement for uniformity. In many areas, school construction has begun to break away from the conventional egg-crate or box design. Change has been slow but is gaining momentum. Not only are facilities now being designed to implement the curriculum of a school, but also schools are becoming more compact and functional. New materials are being developed to further aid construction of school buildings. Much of the furnishings and equipment will also be of some synthetic material.

In attempting to hold down construction costs and to shorten the time it takes to build, the systems approach to construction will become more common. Larger and larger parts, already assembled, will arrive at the construction site for installation. Productivity of labor is increasing faster in factories and other areas thus saving time for construction. Automation will replace much of the menial type of work. A wide variety of materials and components will give designers needed selectivity for individual schools.

Fast track planning and designing will reduce preparation time for building plans thus saving money for the owner. Computer programming and scheduling for the contractor will grow and will include planning certain elements, as well as in ordering and placing equipment needed in a good school facility.

By utilizing acoustical materials, noise (unwanted sound) will be controlled. Reducing the noise level of a school is of utmost importance due to the fact that any group of human beings has a tendency to be noisy. Any material which may be used to help control sound is a valuable asset to a school facility. Because the world is becoming increasingly noisy, greater attention will be given to the control of sound in a school. Widespread use of carpet or soft floor covering will become more common. This kind of floor covering eliminates much of the impact noise and up to sixty per cent of the airborne noise. By using soft floor covering on hard surfaces sound waves are controlled to a considerable degree.

The advent of the open plan school arrangement for classrooms means that sound control in the spaces for teaching is needed more than ever before. There has never been good control of sound, perhaps due to the fact that there may have been more interest in building a facility so it could be maintained easily. There is no way to make the open plan classroom work properly without adequate sound control.

To maintain a constant level of lighting in the classroom, artificial lighting is utilized. Most of today's schools are reducing the number of windows which means less natural light. Schools will continue to reduce or eliminate much of the glass in outside walls. Most of the state regulations have been revised so that the amount of glass in a classroom has been reduced. Artificial light will be controlled in such a way that it will provide better lighting than is the case for natural light. Artificial light is easier to control, especially from the standpoint of brightness difference. Light may not only come from the ceiling but also come from the walls or a permanent fixture, or heat from lights may be channeled in such a way as to help heat the building. Some people believe that incorporating all of the environmental control factors in a school will be expensive. This is not necessarily true. If the facility is designed so that perimeter is reduced, the dollars spent for construction may be reduced.

With the expanded use of soft floor covering, certain types of furniture may be used less in the future. This will be especially true in the lower grades. Furniture for tomorrow will be of a different design from that of today. Hard, uncomfortable chairs will no longer be around. Seating, work tables and other furniture will be designed in such a way as to provide maximum utilization and comfort. All types of furniture will be more colorful, free of safety hazards and will last longer. Desks, tables and chairs will be made of lightweight material that will wear well, be easy to move around and stackable.

The schools of tomorrow will have a better environment for learning as a result of improvements in temperature control. Air-conditioning equipment is now available that will provide for heating or cooling using only 60 seconds for the change over. A temperature of 75 degrees plus or minus two degrees must be maintained. A school which does not provide for proper temperature control will provide a barrier to learning. Without qualification, all schools of the future must be able to maintain an optimum temperature fifty-two weeks a year if there is to be top performance on the part of teachers and pupils. Any school planned today that does not have the capability for a precisely controlled environment is outdated the day that pupils and teachers move into the building.

Systems which are presently being utilized to heat and cool buildings will be vastly improved. Unit ventilators and other types of equipment will give way to some highly sensitive type heating and cooling system. By using such a system, instant heating or cooling may be provided within one minutes whatever the season of the year may be. Thermal control equipment will become more efficient as absorption techniques are improved. In the near future there will be little if any energy loss in exhausted air. Such small quantities of heat may be stored or solar energy may constitute the entire source for heating.

Technological advances do not represent an abrupt break with history but rather a continuation of the steady stream of discoveries, improvements, inventions, and new techniques which have affected education since the invention of the printing press. It is difficult to find a single facet of society which has been immune to the effects of technological change. Machines will be increasingly utilized as an aid to instruction just as textbooks and reference books are used at the present time. Machines will assist the teacher in many routine tasks such as drill or review. Through the use of computer assisted instruction, taped lessons, filmstrips, slides, programmed instruction, and other video tape procedures, teachers may be relieved of much repetitious work that may be done better through the use of the machine. Machines will never take the place of the teacher, but will assist the teacher in doing a better job of teaching. For many years, self-instruction devices and teaching machines have been available to help the teacher do a better job of teaching. Certain techniques may be grouped under the name of programmed instruction. As we know it today, programmed instruction provides for instruction, questions, answers, and other information based upon the individual's capabilities. With the trend toward more and varied machines it is desirable that the building be designed to take advantage of the best equipment available. Planning for teaching machines and others in the initial stage, represents a small per cent of the cost of providing such needs after the building is completed.

Modern scociety is demanding change in the social structure of the nation. This change is being felt by educators at all levels. Such influence will continue to exert pressure on existing organizational patterns demanding improved programs. Modern trends have already brought about changes that appear in urban planning and consolidation of school districts throughout the country. Most every conceivable type of shared services will appear on the scene. Shared services will cut across school district lines. Large school districts made up of many smaller districts will come into being. Services here-to-fore considered unobtainable for small schools have and will become more of a reality. Multimillion dollar facilities housing total service centers will more and more become a part of good planning. Changes in the educational patterns will exceed the forecasts of the visionaries of today.

One such facility already in operation is the John F. Kennedy School and Community Center in Atlanta, Georgia. The Center was designed and built for those who suffer from the blight and deterioration of the inner city. Opened for operation early in 1971, the Center has already become a symbol of hope and inspiration for the people it serves and for many others who are concerned with the hardships of poverty.

Located on a school-park site, near an existing neighborhood health clinic, the John F. Kennedy School and Community Center houses a middle school, and other programs and services provided by more than a dozen agencies, Federal, State, County, City and Private.

The programs and services are operated simultaneously on a year-around basis. Some spaces are shared, but most agencies occupy their own space. While the middle school (grades 6-8) occupies an area of approximately 100,000 square feet, the community facilities, including agency offices, recreation services, and workshop areas, utilize an additional 125,000 square feet.

The community education concept is growing and will be implemented in a variety of ways in the future. School programs will accept and use the total community's resources as a means of accomplishing educational goals. Education will assume an individualized approach and will break away from the notion that learning must take place inside the four walls of a school building.

Tomorrow's school will see more and more opening up of spaces inside the facility. Modularization which at times has led to standardization will lean toward patterns which will add to flexibility of spaces. Many schools are providing open space by leaving out permanent walls. Schools of the future will continue to do this even to a greater extent. In order to have maximum flexibility, school facilities of the future will move more and more toward one large open space such as the loft plan. This large open space will be divided up with visual dividers. Instructional areas which have resisted the open concept will yield. Administrative and guidance areas will be planned with visual dividers or movable partitions. Vocational areas will be divided into high noise level areas and low noise level areas. Within most of the areas will be movable partitions or visual dividers which will serve as visual and sound barriers.

Parents and laymen in their quest for better and more effective utilization of school facilities are insisting upon greater utilization of school buildings as well as using the school building twelve months each year. The expansion of school services through greater utilization of existing spaces will bring about a more effective program. Day care centers will be located in schools in the industrial areas. The program will reach out to include the three- and four-year-olds as well as grades thirteen and fourteen. Schools will assume greater responsibility for recreational programs and educational leaders will provide for a more effective adult program. There will be programs to train adults in varied skills and re-train the unemployed to meet the needs of a dynamic society.

Urban schools can expect to utilize more high rise or joint occupancy buildings in the high cost real estate areas. Schools may spring up over railroads or over subway yards and will span the lower structures. Buildings will be built above and below ground. The role of the school in the ghetto area will require buildings to be placed in densely populated areas with easy access to the school for both children and adults. Entrances will be spread over several blocks with elevated, protected walkways which will lead to the teaching spaces. In such densely settled areas, high rise schools and shared spaces will be utilized. Schools will operate within existing but modernized structures with outdoor recreation areas adjacent to the building or utilize decks at the side of or on top of the facility. The movement toward greater centralization will bring students together by the thousands, using the school within a school, which is sometimes referred to as the educational park. These modern centers will provide an almost unlimited number of services. This will call for new modes of transportation or drastic alteration of already existing systems to fulfill new requirements.

MAINTENANCE

The physical facilities of a school system represent an investment amounting to millions of dollars. Maintaining these facilities is one of the task areas of the chief school administrator as well as the entire staff in the school system. If the administrator overlooks this responsibility, valuable property will deteriorate quickly, leading to large expenditures for repairing the buildings and equipment at a later date. Maintenance that is below par is very costly and may be wasting the taxpayers dollars. A better job is done in maintaining a school building today than in the past, however, there is ample room for improvement. School buildings cost twice as much as was the case ten to fifteen years ago. When budget time comes along, the amount budgeted for maintenance is most always short in that personnel are employed and paid the lowest salary of anyone in the school and very little money is available for supplies and equipment. All of these factors contribute to the kind of maintenance conditions that are undesirable and sometimes unbelievable.

Recent advances in school planning has resulted in the kind of a facility that is functional and innovative, using a variety of materials and equipment. These changes represent important improvements, especially on the inside of a building. Modern practices utilized in planning a school demand personnel with greater technical skills, an understanding of sanitation and general education as well as a better understanding of the role of the school custodian.

Educational leaders are faced with a new set of problems related to custodial care in school plant operation and repairs. Custodians need greater competence in order to bring about better solutions in the solving of problems related to the care of the school plant. A program of in-service education for maintenance workers which deals with adequate equipment and cleaning materials is needed. This program should be designed to meet new custodial needs.

It is an accepted fact that every effort must be made to protect the health of teachers and pupils. Attention given to the total school environment supports the educational program and at the same time makes a contribution to the health of pupils by providing a satisfactory environment for health and safety. A schoolroom that is full of dirt and filth will be full of bacteria. A clean environment develops a cheerful outlook on the part of pupils and teachers. This cheerful outlook promotes cleanliness in the building as pupils really prefer clean spaces. Pupils, teachers and principals know how important a custodian or repairman is in relation to the educational program. A clean, well-kept building with proper heating and cooling has a tremendous influence on the educational program.

The safety and health of teachers and pupils are a part of the responsibility of the head administrator and the principal of a school. This responsibility must be carried out with well-trained custodians who have proper tools and equipment as well as cleaning supplies. Sanitary conditions are achieved through careful school plant management, control of temperature, plus attention to safety. Proper care of the school plant protects teachers and pupils and other personnel against fires and accidents. Many items of equipment on the playground or in the building may lead to accidents. Slippery walks or loose glass can cause bad accidents. Storage of oily or dirty rags or improper wiring in a building may cause fires. The maintenance worker must remove or get rid of all hazards to the safety of pupils and teachers.

Provision for healthful conditions in a schoolroom is one of the most important responsibilities of the principal. A clean building will protect the health of all occupants and also help in developing desirable health habits. Good health also depends to a considerable extent upon proper regulation of temperature, a clean classroom and a good lighting system within the building. Restrooms and other areas must be supervised carefully if the building is to provide a healthful environment.

Many school systems fail to provide for proper maintenance of a school building. This is not as much the fault of maintenance workers as the failure of boards of education and administrators to finance the maintenance program. Proper care seems to be last on the list when it comes to financial support. The most important part of a good maintenance program is the maintenance worker. It is almost useless to try to develop a good program in school plant care when such small amounts are spent for the personnel that is responsible for carrying out the program. Great steps have been taken in recent years toward a better maintenance program yet there is a considerable distance to travel if a good job is to be achieved. It is good to have a good, well-planned school facility with the best of furniture and equipment plus a precisely controlled environment but it is very bad to neglect the maintenance of that facility. If there is poor care of a facility, the educational program will suffer, no matter what the teacher-pupil-ratio may be or how well the building is planned. Carpet, a new floor covering, will not perform the function of absorbing sound waves, and will wear out in three or four years if allowed to become dirty, forming a layer of dirt and filth on top of the carpet fiber. The same thing is true for the school building as a whole and for everything that goes inside the facility. A school building must be maintained in the proper manner.

The following suggestions relate to the responsibilities of the maintenance worker, who is one of the most important employees in a school system.

1. He is responsible for the care of costly property.
2. He is partly responsible for safety of pupils and teachers.
3. He is responsible, in part, for the health of pupils and teachers.
4. He is responsible for maintaining standards of neatness.
5. He is responsible, in part, for providing conditions that tend to make a better environment for learning.
6. He is responsible for developing good will for the institution where he works.
7. He is responsible for repairs to a building.
8. He is responsible through the guidance and assistance of the administrator for bringing about economics in operation.

One of the most pressing problems related to new or older school buildings is providing an adequate program of maintenance. Even though a much better job is being done in providing for proper care of all school facilities, maintenance conditions are deplorable in far too many school buildings. It makes little difference what a facility may cost, it is a waste of money provided for public education, when buildings are allowed to become dirty. The investment in buildings must be protected through an adequate maintenance program. More and more highly sophisticated equipment is used in the modern school, such as electronic aids, audio-visual aids, educational television, scientifically designed furniture and others, need to be kept in good working condition. This equipment will not work as it should unless it is cared for in the proper manner. Teaching spaces, carpet, acoustical tile, or other parts of a school building will not last nor perform the proper function unless maintained properly.

SELECTED REFERENCES

Adams, V. A. "New Lives for Old Schools." *School Management* 14 (January 1970): 4.

"Administrator's Guide to Air Conditioning Basics." *Nation's Schools* 83 (June 1969): 94.

Alanson, Brainard D., and Geodeken, A. E. *Handbook for School Custodians.* Lincoln: University of Nebraska Press, 1961.

American Association of School Administrators. *Planning American School Buildings.* Washington, D.C.: The Association, 1960.

American Association of School Administrators. *Schools for America.* Washington, D.C.: The Association, 1967.

American Association of School Administrators. *Open Space Schools.* Washington, D.C.: The Association, 1971.

American Carpet Institute. *Sound Conditioning with Carpet.* New York: American Carpet Institute, Inc., n.d.

Bailey, William M. "Get the Most from Your Heating Dollar." *Building Operating Management* 17 (June 1970): 84.

Birren, Faber. *Color Psychology and Color Therapy.* New York: University Books, Inc., 1961.

Birren, Faber. *Light, Color and Environment.* New York: Reinhold Book Corporation, 1969.

Bisnow, Mark. "Grounds Care Operations Made Easy." *American School and University* 40 (March 1968): 24-26.

Blake, Peter. *God's Own Junkyard: The Planned Deterioration of America's Landscape.* New York: Holt, Rinehart and Winston, Inc., 1964.

Burris-Meyer, Harold, and Goodfriend, Lewis S. *Acoustics for the Architect.* New York: Reinhold Publishing Corporation, 1957.

Campbell, Edward A. "Flooring—A Vital Environmental Element." *American School Board Journal* 151 (December 1965): 13-14.

"Carpeting in the Schoolhouse." *Overview* 30 (March 1962): 54-56.

"Carpeting Keeps This Open Plan School Quiet." *American School and University* 42 (November 1969): 27.

Castaldi, Basil. *Creative Planning of Educational Facilities.* Chicago: Rand McNally & Company, 1969.

Chambers, James A. "A Study of Attitudes and Feelings Toward Windowless Classrooms." Doctoral dissertation, The University of Tennessee, Knoxville, 1963.

Davini, William C. "Colors for School Interiors." *American School and University,* Vol. 24, No. 4 (1952-53).

Davis, J. T. "Curriculum-Expanding Facilities." *Science Teacher* 36 (February 1969): 26-27.

De Vries, William, and Devenyi, Denes. "Plant Maintenance System Goes Computer." *American School and University* 42 (March 1970): 43.

"Douglas County Includes Carpeting in Renovation of Their 14 Schools." *American School and University* 42 (October 1969): 23.

Educational Facilities Laboratories. *The Cost of a Schoolhouse*. New York: Educational Facilities Laboratories, Inc., 1960.

Finley, E. D. "Growing School Control: Its Growing Heat Costs." *American School and University* 40 (January 1968): 32.

Fiscalini, F. "Who Says We Can't Afford Carpeting?" *School Management* 12 (October 1968): 130-131.

Fitzroy, Daniel, and Reid, John Lyon. *Acoustical Environment of School Buildings*. New York: Educational Facilities Laboratories, Inc., 1963.

Gardner, J. C. "Carpeting is Down to Stay." *American School and University* 41 (November 1968): 52.

Gardner, J. C. "Maintaining Your Carpeting." *American School and University* 42 (May 1970): 17, 50.

Gardner, J. C. "Preparing, Presenting and Controlling the B & G Budget." *American School and University* 40 (April 1969): 69.

Gardner, J. C. "On the Spot Carpet Care." *American School and University* 40 (February 1968): 42-44.

Gilliland, J. W. "How to Guarantee Long Life for Your Carpeting." *American School and University* 41 (September 1968): 51-52.

Gilliland, John W. "The Trend Toward Functional Schools." *American School and University*, Vol. 39, No. 7 (March 1967).

Gilliland, John W. "What Makes A Good Schoolhouse?" *American School and University*, Vol. 36, No. 8 (April 1964).

Harrison, Clyde Barker. "Factors Affecting the Cost of Maintaining Floor Coverings in School Facilities." Doctoral dissertation, The University of Tennessee, Knoxville, 1968.

Hart, A. L. "Safety, Security and Beauty Through Outdoor Lighting." *American School and University* 40 (May 1968): 54-56.

Herrick, John H. "Contract Cleaners Plan Sweep Through Schools." *Nation's Schools* 84 (November 1969): 74-80.

Hickman, L. C. "How Much Lighting Is Enough," *Nation's Schools* 85 (January 1970): 86-88.

Hill, Fred. "New Lighting Schemes For Integrated Designs." *American School and University* 40 (January 1968): 44-46.

"How to Clean Your Floors Faster, Cheaper and Better." *School Management* 10 (December 1966): 7-10.

Lieff, Morris. "Some Structural Facts About Cement-Fiber Roof Decks." *American School and University* 42 (March 1970): 52-54.

Mathie, Alton J. "Scale Build Up Costs in Water-Cooled Air Conditioning Equipment." *Building Maintenance and Modernization* 16 (December 1969): 18.

Maynard, Herbert. "Removing Heat From Light For Best Effects." *American School and University* 41 (January 1969): 42-44.

Maynard, Herbert. "Ruling the Roofs in Rochester Keeps Repair Costs Under Control." *American School and University* 41 (March 1969): 41-42.

McElrath, Robert L. "Instructional Closed-Circuit Television Systems." Doctoral dissertation, The University of Tennessee, Knoxville, 1968.

McQuade, Walter, ed. *Schoolhouse*. New York: Simon and Schuster, 1958.

Mincy, Homer Franklin, Jr. "A Study of Factors Involved in Establishing a Satisfactory Thermal Environment in the Classroom." Doctoral dissertation, The University of Tennessee, Knoxville, 1961.

National Commission on Technology. "Automation, and Economic Progress." *Statements Relating to the Impact of Technological Change*. Washington, D.C.: The Commission, 1966.

Neutra, Richard. *Survival Through Design*. New York: Oxford University Press, 1954.

"Non-Electric Thermostatic Zone Control." *Building Maintenance and Modernization* 16 (August 1969): 44.

Phillips, Paul J. "Reactions of Students and Teachers to Carpeted Teaching Spaces." Doctoral dissertation, The University of Tennessee, Knoxville, 1968.

"Planned Lighting Helps Learning." *Modern Schools* (March 1969): 6-7.

"Products." *American School and University* 42 (December 1969): 30-38.

"Remodeling Project at Chicago School to Help Curb Vandalism." *American School and University* 41 (September 1968): 46.

Roaden, Ova Paul. "The Essential Elements of Educational Specifications for School Plant Facilities." Doctoral dissertation, The University of Tennessee, Knoxville, 1963.

Sampson, Foster K. "How Much Lighting Is Enough." *Nation's Schools* 85 (January 1970): 86-88.

School Government Publishing Co. *School Buildings Equipment and Supplier* 8 (October 1969): 18-19.

Sheard, Charles. *The American Journal of Optometry*, July 1936. Quoted in *The Schoolhouse*, edited by Walter McQuade. 1958.

Skidmore, C. E. "Fistful of Building Ideas You Can Use." *School Management* 12 (September 1968): 78-81.

Sparks, Harry M. *School Plant Maintenance and Custodial Services*. Frankfort: Division of Statistical Services, State Dept. of Education, 1967.

Thomas, J. E. "Designed for Teaching." *School and Community* 55 (February 1969): 14.

Tonigan, F. Richard. "What Plant Management Can Do." *School Management* 14 (January 1970): 16.

"Vinyl Floor Coverings Need Pampered Upkeep Program." *Nation's Schools* 83 (May 1969): 100-102.

Winslow, C. E. A., and Herrington, L. P. *Temperature and Human Life*. Princeton, New Jersey: Princeton University Press, 1949.

Wintering, Terrance. "Coating Selections." *Building Operating Management* 17 (March 1970): 38.

Womack, Darwin Wasson. "A Study of Factors Involved in Establishing a Satisfactory Acoustical Environment in the Classroom." Doctoral dissertation, The University of Tennessee, Knoxville, 1962.

Yates, Donald P. "Flexibility in School Plant Development and Utilization." Doctoral dissertation, The University of Tennessee, Knoxville, 1968.

Index

AASA, 10, 80
Accountability, 67, 131
Adams, Dewey A., 76
Administration, definition of, 27
Administrative innovation and the supporting technology, 154
Administrative internship, 27
Administrative organizations,
  the future of educational, 71
  the nature of in educational institutions, 68
Administrative team,
  climate essential for an, 83
  human needs must be met in an, 87
  a "two-stream" model of an, 93
Administrative team concept, 79
  decision making in an, 89
Administrative team leadership, 99
Administrative team organizations,
  implication for the superintendent, 92
  some critical factors in the success of, 97
Administrative team operation,
  conditions essential for, 81
  creativity within an, 91
  open communication as a requisite for, 85
  principles of, 80
Administrative theory, 26
Administrative theory as a process, 40
Alam, Dale, 83
Allen, Dwight, 163
American dream, the, 1
American proposition, the, 9, 10
American schools, the future shape of, 11
ASCD, 10, 141
Atwell, Charles, 76

Barnard, Chester I., 33, 100
Bates, D. M., 183
Berla, David, 86
Bestor, Arthur E., 72
Bidwell, Charles E., 183
BSCS, 165
Budget preparation activities, 227
Buffington, Reed L., 45
Building specifications, 270
Burns, Richard W., 107
Butler, Thomas M., 183

Callahan, Raymond, 100, 192
Campbell, Ronald F., 173
Carlson, Richard O., 151
Center for the Study of Public Policy, 219
Central Purpose of American Education, the, 10
CHEM STUDY, 164
Chowdry, Kamla, 52
Clark, D. L., 183
Classrooms without walls, 156
Client system, 19, 241
Coladarci, Arthur P., 39
Combs, Arthur W., 84
Commager, Henry Steele, 8
Committee for Economic Development, 140, 144
Communication, a simple model, 185
Communications in organizations, defined, 85
Comprehensive school, the need for a, 135
CPM, 3, 225
Cunningham, Luvern, 2, 16, 17, 20
Curricular change, systematizing, 136
Curricular guidelines, general, 129

Curricular problems, 108
  the drug culture, 111
  environmental education, 116
  race and ethnic relationships, 113
  the welfare crisis, 108
Curricular programs, with high priorities,
  adult and continuing education, 126
  career education, 121
  community college education, 123
  compensatory programs, 119
  early childhood education, 118
  education for exceptional children, 119
Curriculum development and administration,
  future challenges of, 132
Curriculum implications of competency based education, 131

Davis, Daniel, 38
Decision making process, 33, 61
Demeke, Howard J., 25
Democracy, need for enlightened electorate, 8, 10
Desegregation of schools, 78
Dewey, John, 23, 210, 258
Differentiated staffing, 188
Drug abuse, 111
Dynamic conservatism, 139

Eastburn, David P., 206
Economic man vs. social man, 206
Education,
  hopes for its usefulness in the United States, 12
  significant changes and their implications, 16
  test of its effectiveness, 12
  the cornerstone of freedom in the United States, 11
  the need for relevance, 134
  the promises of, 21
  the purposes of, 11
Educational administration,
  the major functions of, 61
  role theory as a dimension of, 40
  theory in, 25
Educational administrator, the role of as a catalyst, 45

Educational expenditures and the GNP, 211
Educational finance disparities, capacity, and effort, 216
EDUCATIONAL FUTURISM IN 1985, 76
Educational planner, the, 64
Educational planning, 234
Educational Policies Commission, 118
EDUCATIONAL TURNKEY SYSTEMS, 229
Education for citizenship, 8
EDUCATION AND THE CULT OF EFFICIENCY, 100, 192
Education as an economic problem, 205
Eight State Project, 16
Environmental education, 116
Evaluation of instruction, 168
Evaluation within and of the educational organization, 66

Fagon, R. E., 28
Fantini, Mario D., 77
Federalism in transition, 13, 18
Finance, micro systems—local resource management, 221
Fiscal capacity and equality of educational opportunity, 217
Flexible scheduling, 156
Flexible space, 258
Forecasting technology, 201
Foster, Richard, 141
Free public schools, 8
Fully-functioning people, 88

Galbraith, John Kenneth, 207
Garner, Paul L., 126
Garvue, Robert J., 200
General systems theory, 3, 8, 19, 25
  the promise of, 20
Getzels, Jacob W., 39
Gilliland, John W., 256
Glasser, William, 77, 140
Goodlad, John I., 8
Gouldner, Alvin W., 176
Governance of schools, a central issue in education, 15
Granger, Robert L., 21
Griffiths, Daniel E., 28, 37, 89, 173

Gross, Donald, and Gross, Beatrice, 77
Gross National Product (GNP), 209
Group dynamics, 160

Hall, A. D., 28
Halpin, Andrew, 25, 32, 46, 64, 81
Hansen, W. Lee, 234
Havighurst, Robert L., 79
Hawaii, no local school districts, 12
Hawthorne effect, 142
Hawthorne study, 142
Headstart, 109, 208
Herzberg, F., 183
HEW, 18
Holder, Lyal E., 60
Holt, John, 140
Human needs, an important consideration in personnel administration, 176
HUMAN RELATIONS IN SCHOOL ADMINISTRATION, 38

Implementation as a process in administration, 65
IMPLICATIONS FOR EDUCATION OF PROSPECTIVE CHANGES IN SOCIETY, 17
Innovation, 141
  the administrator's role in, 159
  the conditions for, 159
  strategies in bringing about, 163
Innovation and the innovator, facilitating the, 167
Innovations, a survey of, 145
  computer assisted instruction (CAI), 151
  individually prescribed instruction (IPI), 149
  instructional television, 145
  the new curricula (New Math, BSCS, etc.), 153
  programmed instruction, 149
  team teaching, 147
Institutional system, 18, 19

Jarvis, Oscar T., 107
Jefferson, Thomas, 1, 12
Job Corps, 109
Johansen, John H., 238
Johns, Roe L., 17, 20

Johnson, Eldon D., 183
Jones, J. I., 183

Kalamazoo Case (1872), 258
Katz, Robert L., 38
Kimbrough, Ralph B., 173
Kohl, Herbert, 140
Koontz, Harold, 87

Lancastrian system of education, 257
Lao Tze, 82
Lasswell, Harold D., 61, 65
Leadership,
  definitions of, 28, 82
  new concepts of, 192
  the Ohio State studies, 192
Leslie, Larry L., 138
Levin, Henry M., 218
Likert, Rensis, 81
Linear schools, 157
Lipham, James, 28
Lippitt, Gordon L., 89

McGregor, Douglas, 88
Macro systems, 202
Mager, Robert F., 143
MANAGEMENT AND THE WORKER, 52
Managerial system, 18, 19
Mann, Horace, 22
Marland, Sidney P., 3, 20
Maslow, Abraham H., 87
Maxson, Robert C., 1
Mayo, Elton, 192
Medsker, Leland, 45
Merton, Robert K., 175
Metatheory, 21
Model Cities Project, 20
Moore, Harold E., 182
Myers, Phyllis, 215

NASSP, 10
National Urban Coalition, 215
Newcomb, Theodore M., 52

Objectives, 143
O'Donnell, Cyril, 87
Ohm, Robert E., 76
Ombudsman, 186
Ordiorne, George, 103

Organization, bureaucratic, 174
  definitions of, 30
  types of, 30
Organizational climate, 64, 83
Organizational renewal, 45
Organizations, line and staff, 69, 73, 94
  the ellipse model, 70

Parkinson's law, 159
Parsons, Talcott, 19
Perceptual differences of good and poor helpers, 84
Performance contracting, 157
Personnel administration, 193
  changes in supply and demand, 195
  the changing concept of, 191
  a competence model, 41
  humanism, 194
  new legal relationships, 194
  new race relations, 193
  parameters affecting, 172
Personnel administrator, the role of, 177
Personnel program, components of the school, 179
Planning as a process in administration, 63
Posamanick, Benjamin, 183
Poverty culture, 109
PPBES, 200
PPBS, 3, 21, 225, 228
Presidents' Commission on National Goals, 10
Principles of administration, 32
PRINCIPLES OF SCIENTIFIC MANAGEMENT, THE, 100
Professional negotiations, 194
Purpose of education in the United States, 1, 3, 8

Quincy school, 257

Rettig, Soloman, 183
Rockefeller Foundation, 141
Roethlisberger, F. J., 31

Samuelson, Paul A., 202
Savage, Ralph M., 184
Saxe, Richard N., 33

Schon, Donald A., 139
School building maintenance, 276
School buildings, the evolution of, 257
  innovations and trends, 259
  acoustical control, 264
  air conditioning, 267
  compact buildings, 259
  economy designs, 259
  environment for learning, 262
  flexible furniture and equipment, 269
  humidity control, 264
  media and machines, 261
  soft floors, 266
  wall colors, 268
School buildings of the future, 272
Schools' role in effecting social change, 112
Scientific management, 69
Serrano and educational fiscal reform, 215
Serrano vs. Priest, 213, 217, 236
Seven Cardinal Principles of Education, 10
Silberman, Charles E., 77, 140
Simon, Herbert, 100
Sistrunk, Walter E., 1
Smith, Adam, 20, 219
Social systems, 19, 238
Stewart, L. H., 183
Stogdill, Ralph, 82
Suehr, John H., 183
Supra and sub-systems, 18, 239
System, information, 246
Systems analysis, 224, 242
Systems analyst, the, 64
Systems approach, 243
  a conceptual framework, 35
Systems approach to administration—a suggested model, 47
Systems approach to curricular change, 137
Systems approach with specific learning objectives, 155
Systems architecture, 156
Systems, definitions of, 28, 47
Systems of management, the need for, 31
Systems model for finance, 222

Systems model, a simplified, 245
Systems team, 145

Taylor, Frederick, 100, 192
Teacher involvement in administrative decisions, 9, 10, 14, 72, 77
Team concept of administration is not a theory of administration, 102
Team management, the preparation of administrators for, 100
Technical system, 19
Ten Imperative Needs of Youth, 10
Theoretical framework, lack of operational model, 38
Theoretician, the task of, 45
Theories of administration, 32
Theories of educational administration, the search for, 38
Theory of administrative behavior is not universally accepted, 101
Theory, definition of, 27
Theory in educational administration, 25
Theory must suggest methodology, 46

Theory, the uses of, 56
Thrasher, James M., 60

United States Constitution, 12
  government of schools a state function by inference, 18
University of Georgia Research and Development Center in Early Educational Stimulation, 119
Urbanism, dispelling the myth of, 133

Value Added Tax (VAT), 214
Venn, Grant, 134
Voucher plans, seven alternatives, 220
Vouchers, educational, 219

Walton, John, 33
Weinrich, Ernest, 38
Williams, James O., 172
Womack, Darwin, 256
Worner, Wayne M., 76

Year-round school, the need for a, 135
Young, Milton A., 77